High-Pressure Sulfidation of Hydrotreating Catalysts:

Genesis and Properties of the Active Phase

The research described in this thesis was performed at the section Fundamental Aspects of Material and Energy of the Department of Radiation, Radionuclides and Reactors, Faculty of Applied Science, Delft University of Technology, Mekelweg 15, 2629 JB Delft, The Netherlands

High-Pressure Sulfidation of Hydrotreating Catalysts:

Genesis and Properties of the Active Phase

PROEFSCHRIFT

ter verkrijging van de graad van doctor
aan de Technische Universiteit Delft,
op gezag van de Rector Magnificus prof. dr. ir. J.T. Fokkema,
voorzitter van het College voor Promoties,
in het openbaar te verdedigen
op dinsdag 29 april 2008 om 12.30 uur
door

Achim Iulian DUGULAN

physics and chemical engineer
Universiteit van Boekarest, Roemenië

geboren te Râmnicu Vâlcea, Roemenië

Dit proefschrift is goedgekeurd door de promotoren:
Prof. dr. J.A.R van Veen
Prof. dr. I.M. de Schepper

Samenstelling promotiecommissie:

Rector Magnificus voorzitter	
Prof. dr. J.A.R van Veen	Technische Universiteit Eindhoven, promotor
Prof. dr. I.M. de Schepper	Technische Universiteit Delft, promotor
Prof. dr. R. Prins	Swiss Federal Institute of Technology (ETH) Zürich
Prof. dr. J.A. Moulijn	Technische Universiteit Delft
Prof. dr. B. Weckhuysen	Universiteit Utrecht
Dr. E.J.M. Hensen	Technische Universiteit Eindhoven
Dr. M.W.J. Crajé	Elektriciteits-Produktiemaatschappij Zuid-Nederland (EPZ) Borssele

Dr. E.J.M. Hensen heeft als begeleider in belangrijke mate aan de totstandkoming van het proefschrift bijgedragen.

ISBN 978-1-58603-863-2

Keywords: Hydrodesulfurization; ^{57}Co Mössbauer emission spectroscopy, ^{182}W Mössbauer absorption spectroscopy; high-pressure; sulfide catalysts; Co-Mo-S.

Published and distributed by IOS Press under the imprint of Delft University Press

Publisher
IOS Press
Nieuwe Hemweg 6b
1013 BG Amsterdam
The Netherlands
tel: +31-20-688 3355
fax: +31-20-687 0019
email: info@iospress.nl
www.iospress.nl
www.dupress.nl

PRINTED IN THE NETHERLANDS

Table of Contents

Chapter 1

General introduction

Abstract

Hydrotreating is one of the key catalytic processes in oil refineries. The stringent environmental legislation on transportation fuel quality and the gradually decreasing availability of lighter types of crude oil underline the necessity to further improve the catalytic activity of hydrotreating catalysts. In spite of the progress made in the fundamental understanding of the active phase morphology, the metal-support interaction and reaction mechanisms, many details about the nature and stability of the active sites have not yet been elucidated. One specific issue not widely addressed yet is the influence of the sulfidation pressure on the active phase in hydrotreating catalysts. In industrial practice, catalysts are brought in their active, sulfided form at elevated pressure. The main objective of the present thesis is to understand the effect of the sulfidation pressure on the active phase structure in CoMo and NiW catalysts.

1.1 Hydrotreating

With world energy demand still increasing, oil is expected to remain the primary source of energy around the globe contributing to approximately 40% of the total. The global petroleum demand is projected to increase from about 82 million barrels per day in 2004 to 111 million barrels per day in 2025. The transportation fuels will account for 87% of the increase in oil consumption and for more than half of the total world oil consumption [1].

Important products such as LPG, gasoline, kerosene and diesel oil are obtained in oil refineries by atmospheric distillation of the crude oil. The different boiling fractions are subsequently treated in catalytic processes like hydrotreating, isomerization, alkylation and cracking in order to meet the various specifications of the products. A simplified representation of a typical oil refinery is given in Figure 1.1. Almost all product streams are purified in hydrotreating units, which makes such process one of the most important catalytic processes.

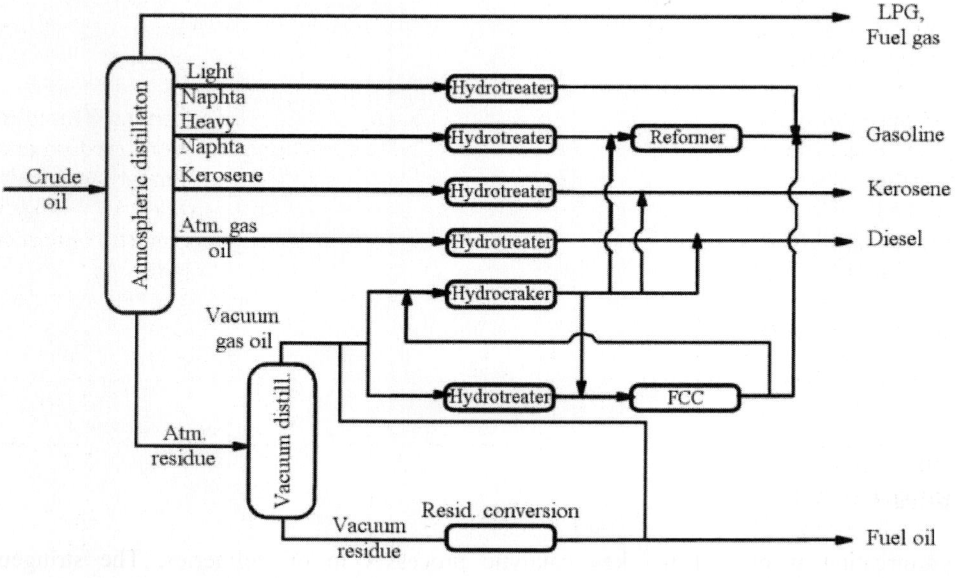

***Figure 1.1** Simplified flow scheme of an oil refinery (adapted from [2]).*

In hydrotreating sulfur, nitrogen, oxygen and metal atoms are removed from the different petroleum streams and unsaturated hydrocarbons (mostly olefins) are saturated [3]. All these processes, hydrodesulfurization (HDS), hydrodenitrogenation (HDN), hydrodeoxygenation (HDO), and hydrodemetallization (HDM), are using hydrogen as a reactant and catalysts based on transition metal sulfides.

The main reasons for refineries to perform hydrotreating are of environmental and economic nature. Besides protection of downstream catalysts from poisoning by sulfur, stringent environmental legislation has been aimed at the reduction of sulfur oxide and emissions from fuel combustion. The full introduction of "zero-sulfur" gasoline and diesel fuels [4] – with less than 10 mg/kg (ppm) sulfur – must be completed by 1 January 2009 in the European Union. Table 1.1 shows the evolution of standards for the amount of sulfur in transportation fuels. The sulfur content in diesel and gasoline should be decreased by a factor of five by 2009 as compared to 2005. These restrictions strongly require the development of new processing technologies and improved HDS catalysts.

Table 1.1 *Evolution of product specifications for transportation fuels (Source: European Parliament)*

	2000	2005	2009
Sulfur (ppm)			
Gasoline	150	50	10
Diesel	350	50	10

The increasing demand for transportation fuels combined with the decreasing demand for fuel oil [1] and the gradually decreasing availability of lighter types of oil means that heavier fractions have to be cracked into lighter ones. One of the most important processes in modern refineries, the fluid catalytic cracking (FCC) is employed to increase the gasoline yield by cracking the heavy molecules at high temperatures using zeolite catalysts [5]. In hydrocracking the acid catalyst is combined with a hydrogenation catalyst. The aim is often to produce diesel fuel, especially in the European Union [6]. Catalytic reforming is used to convert linear hydrocarbons into branched and aromatic compounds by isomerization and cyclization, increasing the octane number of the naphtha feed [7]. The acid catalysts used in FCC and the transition metals applied for catalytic reforming are poisoned by nitrogen compounds [3]. To avoid the rapid deactivation of these catalysts and improve product specifications, sulfur, nitrogen and metal compounds are removed by hydrotreating.

Table 1.2 *Typical properties of various crude oils*

	Arabian Light	Arabian Heavy	Attaka (Indonesia)	Boscan
Sulfur (wt%)	1.8	2.9	0.07	5.2
Nitrogen (wt%)	0.1	0.2	<0.1	0.7
Oxygen (wt%)	<0.1	<0.1	<0.1	<0.1
V (ppm)	18	50	<1	1200
Ni (ppm)	4	16	<1	150
Wt% distilled at 360 °C	54	47	91	20

On the crude oil side, one has to deal also with significant variations in the composition depending on its origin. Some examples of various crude oils are given in Table 1.2 [3]. The quality of the world's crude oil that will be produced in the near future will decrease [8], while refineries are making considerable efforts to run the optimal mix of crudes depending on the desired output products, the refinery's equipment and available catalysts. There is a clear need to further improve the catalytic activity of hydrotreating catalysts to arrive at clean transportation fuels from ever more heavy feedstocks.

1.2 Hydrotreating catalysts

HDS catalysts consist of mixed sulfides of Co or Ni and Mo or W dispersed over a γ-alumina support. The commonly used combinations of active elements depend on the application and the desired activity and selectivity of the catalysts. CoMo sulfide catalysts are preferred for HDS operations, while the NiMo sulfide are very good in HDN and when more intensive hydrogenation is required. The NiW catalyst has the highest activity for hydrogenation of low-sulfur feedstock [9,10], but a drawback is the higher catalyst cost due to the use of W.

To reach the required reduction of the sulfur levels by HDS, it becomes necessary to remove sulfur from compounds that are most difficult to desulfurize. Figure 1.2 shows a qualitative relationship between the type and size of sulfur molecules and their relative reactivities [11]. The HDS reactivity over the same catalyst decreases in the order: thiols and (di-) sulfides > thiophenes > benzothiophenes > dibenzothiophene (DBT) > 4,6-dimethyl dibenzothiophene [12,13]. The molecules containing side chains in position close to the sulfur atom are very difficult to desulfurize because of the steric hindrance imposed by the alkyl groups [14].

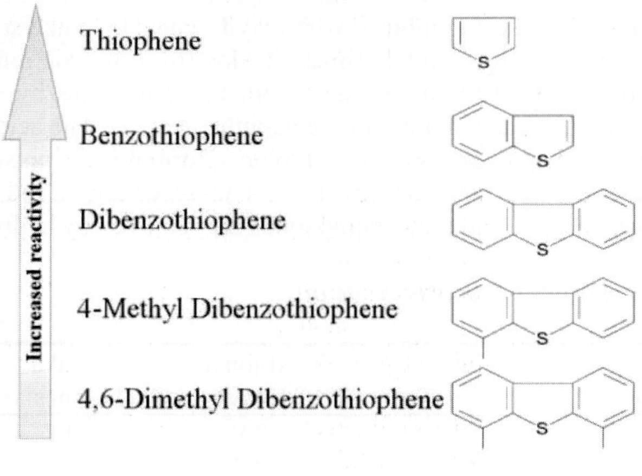

***Figure 1.2** Relative reactivities of various organic sulfur compounds in HDS.*

The HDS of dibenzothiophenes and alkylated dibenzothiophenes proceeds by two main pathways: direct extraction of the sulfur atom from the molecule (hydrogenolysis) and the hydrogenation of the aromatic ring, followed by C-S bond breaking [15], as shown in Figure 1.3.

DBT is desulfurized preferentially via the direct extraction route. When alkyl groups are attached to the aromatic rings close to the sulfur atom, the hydrogenation of one of the rings lifts the steric hindrance and makes the intrinsic reactivity for the HYD pathway higher than for the DDS route [16]. Consequently, catalysts with high selectivity for desulfurization via the hydrogenation route are required for very deep HDS and NiW catalysts are a promising option [17].

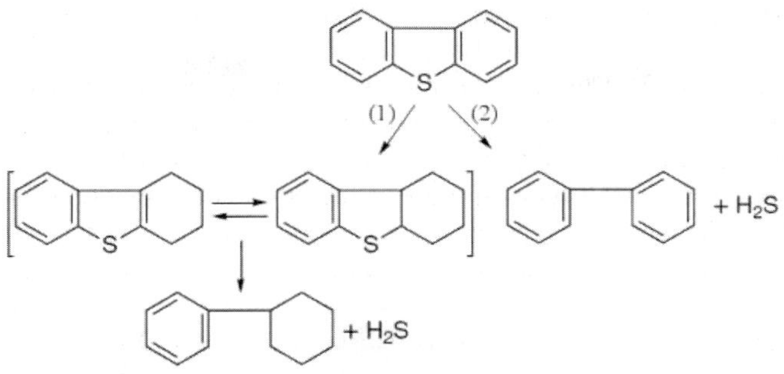

***Figure 1.3** Reaction pathways for the HDS of DBT: (1) hydrogenation (HYD) and (2) direct desulfurization (DDS).*

Important progress has been made towards understanding the structure and nature of the catalytic sites. The evolution of the structural models proposed for the active phase of hydrotreating catalysts has been summarized in many reviews [3, 18–24]. However, the origin of the catalytic synergy between the two main elements of the active sites is still a subject of great debate. Table 1.3 summarizes the main theories and proposals developed to explain this synergy over the last few decades [25].

***Table 1.3** Proposed models of the structure of hydrotreating catalysts*

Authors	Structural model	Reference
Lipsch and Schuit	Monolayer of oxide	[26]
Delmon	Contact synergy	[27-29]
Schuit and Gates	Oxysulfide monolayer	[30]
Voorhoeve and Stuiver	Intercalation of Co	[31]
Farragher and Cossee	Pseudo-intercalation	[32]
Aoshima and Wise	Structural defect	[33]
Jacquin	Mixed sulfide	[34]
Okamoto et al.	Metallic cobalt	[35]
Harris and Chianelli	Electronic effect	[36]
Topsøe	"Co-Mo-S" phase	[37,38]
Delmon	Remote control	[28,29]
Chianelli and Daage	Rim/edge contribution	[39]
Ledoux	Specific configuration of Co	[40]
Prins, de Beer and others	Support effect	[41-43]

Making use for the first time of an *in situ* technique (Mössbauer emission spectroscopy (MES)) for direct measurements of the local structure of Co sulfide species in hydrotreating catalysts, Topsøe and co-workers [37,38] introduced the Co-Mo-S model: Co atoms located at the edges of MoS_2 slabs. A specific MES spectrum, consisting of a doublet with an isomer shift (IS) of 0.22±0.05 mm s^{-1} and a quadrupole splitting (QS) in the range of 1.0 - 1.3 mm s^{-1} was assigned to the Co-Mo-S phase [44].

The Co-Mo-S spectrum observed with MES is different from the spectra of other species that may be present in typical sulfided CoMo catalysts [e.g., bulk Co_9S_8 and Co incorporated in the Al_2O_3 support – for alumina supported catalysts], as shown in Figure 1.4. Detailed structural information about the catalysts under real *in situ* conditions can be obtained using MES as a fingerprint technique, i.e. by systematically comparing the Mössbauer spectra of the catalysts with those of the reference compounds [45].

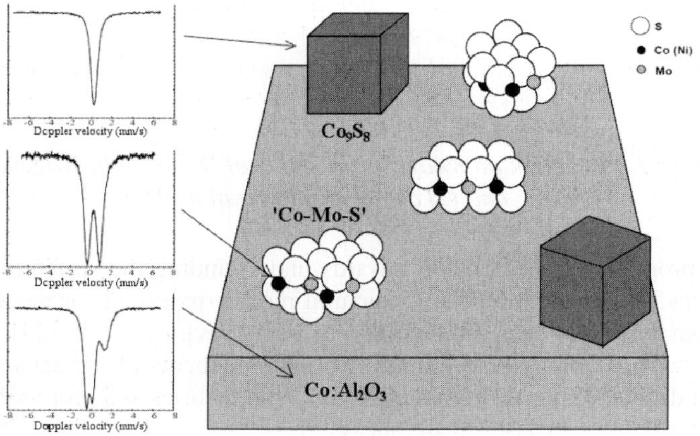

Figure 1.4 *Schematic representation of the different structures present in a sulfided CoMo/Al₂O₃ catalyst, adapted from [3, 46].*

Wivel et al. [47] related the high activity of CoMo catalysts to the amount of Co present in the Co-Mo-S phase. By preparing catalysts with different activities and the same MES signal, it was later shown that no simple relationship exists between the amount of Co-Mo-S and activity [48,49]. Combined MES and extended X-ray absorption fine structure (EXAFS) measurements allowed Crajé et al. [50–53] to improve the Co-Mo-S model by showing that the Co sulfide species located at the MoS_2 edges differ in particle size and/or ordering. It was concluded that the Co sulfide particles in the Co-Mo-S phase, containing only one Co atom, exhibit the largest QS value, with larger Co sulfide species having somewhat smaller QS values [45]. For high Co/Mo ratios and after sulfidation at higher temperatures, crystalline Co_9S_8 particles are formed and the remote control model [27] may apply. According to this model, the activity and selectivity of the catalyst is related to the close interaction of Co_9S_8 and MoS_2, a contact synergy between this phases occurring during the catalytic reaction, the promotor atoms activating and spilling over hydrogen to the MoS_2 edges, increasing the activity of the active sites.

From activity comparison studies [54,55] it was found that high-temperature sulfidation treatment may induce transformation of the Co-Mo-S phase from an incompletely sulfided Type I to a highly active Type II Co-Mo-S, where all Mo-O-Al linkages with the alumina support are sulfided. The use of complexing agents like NTA (nitrilotriacetic acid), which decreases Mo-support interactions [56], can also lead to Type II Co-Mo-S phase formation. NTA also retards the sulfidation of Co to temperatures where MoS_2 particles are already present, favoring Co-Mo-S phase formation [57,58] and induces the development of stacked MoS_2 layers.

The influence of the morphology and orientation of MoS_2 clusters on the catalytic activity has been the subject of many studies [59-61]. The results show that MoS_2 slabs with high aspect ratios (slab thickness divided by slab lateral dimension) have higher intrinsic activities than MoS_2 slabs with low aspect ratios. Two types of active sites on the MoS_2 clusters were proposed in the rim-edge model: the rim sites (located on the top and bottom layers of the slabs) that are responsible for the hydrogenation and the edge sites responsible for both hydrogenation and direct desulfurization [39]. Although multi-layered MoS_2 structures were observed in catalysts containing Type II Co-Mo-S, they may not be a necessary prerequisite for the formation of Type II structures, but just a secondary effect of the weak support interactions [62]. High-angle annular dark-field scanning transmission electron microscopy (HAADF-STEM) studies indicate that the active Type II species can be present as single-layers [63]. Recent scanning tunneling microscopy (STM) observations [64] and also density functional theory (DFT) calculations [65] have further revealed that the Co atoms are substituted into the edge structure of MoS_2 nanoclusters and prefer to be located at the S edges of MoS_2.

For NiW catalysts a Ni-W-S structure similar to the Co-Mo-S phase was observed [66,67]. Type I and Type II Ni-W-S species were identified also in these catalysts [68,69], with a low sulfidation temperature phase (Type I) having high HYD activity and a high sulfidation temperature phase (Type II) with high HDS activity [68]. Because of the formation of strong W-O-Al bonds with the alumina support [70], NiW/Al_2O_3 catalysts are more difficult to sulfide as compared to Mo based catalysts, making it possible to study intermediate stages of sulfidation as separate, stable phases [71].

In spite of the progress made in our fundamental understanding of the active phase morphology, the support interaction and reaction mechanisms, many details about the nature and stability of the active sites have not yet been elucidated, especially for catalysts activated under conditions relevant to industrial catalysis.

1.3 Aim of the research

The subject of this study is a systematic investigation of the effect of the sulfidation pressure on the active phase structure in CoMo-based hydrotreating catalysts using ^{57}Co Mössbauer emission spectroscopy. ^{182}W Mössbauer absorption spectroscopy for W-containing solids has been implemented and the first-time report of ^{182}W Mössbauer spectra on the sulfidation of supported (Ni)W catalysts is presented. Complementary information has been obtained from extended X-ray absorption fine structure (EXAFS) and high-resolution transmission electron microscopy (HREM) measurements. An important step in the formulation of design criteria is the establishment of structure-activity correlations which aims to relate catalytic performance to specific catalyst properties. Gas-phase dibenzothiophene HDS activity tests were performed in collaboration with Eindhoven University of Technology in order to determine such correlations. Although considerable differences were observed between the activity of mixed sulfide catalysts in liquid phase, more frequently employed in industrial practice, and gas-phase HDS reactions, catalysts with similar surface morphologies show a similar pattern when comparing their activity in the two reaction modes [69].

In this work, the effect of the sulfidation pressure on the active phase structure in CoMo catalysts is investigated. To this end, the sulfidation of supported CoMo catalysts was followed as a function of temperature by ^{57}Co MES, Mo and Co K-edge EXAFS and TEM in a systematic manner. Although it is not straightforward to distinguish MES spectra of the Co-sulfide species in catalysts containing only Co and CoMo catalysts [21,52], an attempt is made to identify a sulfidation pattern that characteristically occurs when optimal Co-Mo-S structures are formed.

For hydrotreating catalysts, it is known that the morphology of the active phase can change depending on the reaction conditions. As important changes in the catalyst structure may occur as a function of temperature and pressure [22, 72,73], it is important to perform characterization studies under temperature and pressure conditions relevant to industrial catalysis. In the present thesis, the first in situ high-pressure MES study of the sulfidation of CoMo hydrodesulfurization catalysts is reported. In industrial practice, sulfidation is generally carried out at elevated pressures. One expects higher sulfidation rates as compared to atmospheric pressure sulfidation which has been the general method for sulfidation in most academic research. The first-time application of ^{182}W Mössbauer spectroscopy for characterization studies on catalysts is presented.

In chapter 2, a brief introduction of the applied experimental techniques is given. The evolution of the active phase as a function of sulfidation pressure and temperature is studied for calcined CoMo/Al_2O_3 (chapter 3), for uncalcined CoMo/Al_2O_3 (chapter 4), for carbon-supported CoMo catalysts (chapter 5) and for CoMo/Al_2O_3 catalysts promoted by NTA and phosphate (chapter 6). In chapter 7, structure-activity relations are derived, especially aimed to explain the effect of high-pressure sulfidation on the active phase composition and HDS activity. Finally, chapter 8 addresses ^{182}W Mössbauer spectroscopy applied to study the sulfidation process of NiW/Al_2O_3 and NiW/ASA catalysts.

References

1. Energy Information Administration, *"Annual Energy Outlook"*, DOE/EIA-0383, Washington, 2006.
2. J.W. Gosselink, Cattech 4 (1998) 127.
3. H. Topsøe, B.S. Clausen, F.E. Massoth, in: J.R. Anderson, M. Boudart (Eds.), Hydrotreating Catalysts, in: Catal. Science Technol., vol. 11, Springer, Berlin, 1996.
4. Directive 2003/17/EC of the European Parliament and of the Council on the Quality of Petrol and Diesel Fuels, Brussels, 3 March 2003.
5. J.G. Speight, Catal. Today 98, 2004, 55.
6. J.A.R. van Veen, in: M. Guisnet, J.P. Gilson (Eds.), Zeolites for Cleaner Technologies, Imperial College Press, London, 2002, p. 131.
7. C. Marcilly, J. Catal. 216, 2003, 47.
8. R. Paley, M. Pilling, World Pumps 2006, 2006,24.
9. B.H. Cooper, A. Stanislaus, P.N. Hannerup, Hydrocarbon Processing, 72 (1993) 83.
10. H. Yasuda, M. Higo, S. Yoshitomi, T. Sato, M. Imamura, N. Matsubayashi, H. Shimada, A. Nishijima, Y. Yoshimura, Catalysis Today, 39 (1997) 77.
11. K.G. Knudsen, B.H. Cooper, H. Topsøe, Appl. Cat. A 189 (1999) 205.

12. M.J. Girgis, B.C. Gates, Ind. Eng. Chem. 30 (1991) 2021.
13. P.T. Vasudevan, J.L.G. Fierro, Catal. Rev.-Sci. Eng. 38 (1996) 161.
14. X. Ma, K. Sakanishi, I. Mochida, Ind. Eng. Chem. 33 (1994) 218.
15. M.L. Vrinat, Appl. Catal. 6 (1983) 137.
16. T. Kabe, A. Ishihara, Q. Zhang, Appl. Cat. A 97 (1993) L1.
17. H.R. Reinhoudt, R. Troost, A.D. van Langeveld, J.A.R. van Veen, S.T. Sie, J.A. Moulijn, J. Cat. 203 (2001) 509.
18. G.C.A. Schuit, B.C. Gates, AIChE J. 19 (1973) 417.
19. R. Prins, V.H.J. de Beer, G.A. Somorjai, Catal. Rev. Sci. Eng. 31 (1989) 1.
20. B. Delmon, Catal. Lett. 22 (1993) 1.
21. M.W.J. Crajé, V.H.J. de Beer, J.A.R. van Veen, A.M. van der Kraan, in: M.L. Occelli, R.R. Chianelli (Eds.), Hydrotreating Technology for Pollution Control, Dekker, New York, 1996, p. 95.
22. S. Eijsbouts, Appl. Catal. A 158 (1997) 53.
23. D. Whitehurst, T. Isoda, I. Mochida, Adv. Catal. 42 (1998) 345.
24. J.W. Niemantsverdriet, Spectroscopy in Catalysis, Wiley–VCH, 2000, p. 233.
25. P. Grange, X. Vanhaeren, Catal. Today 36, 1997, 375.
26. J.M.J.G. Lipsch, G.C.A. Schuit, J. Catal. 15 (1969) 179.
27. B. Delmon, Bull. Soc. Chim. Belg. 88 (1979) 279.
28. B. Delmon, React. Kinet. Catal. Lett. 13 (1980) 203.
29. B. Delmon, Appl. Catal. A 113 (1994) 121.
30. G.C.A. Schuit, B.C. Gates, AIChE J. 19 (1973) 417.
31. R.J.H. Voorhoeve, J.C.M. Stuiver, J. Catal. 23 (1971) 243.
32. A.L. Farragher and P. Cossee, in: J.W. Hightower, Editor, Prof. 5th I.C.C., North Holland, Amsterdam (1973), 1301.
33. A. Aoshima, H. Wise, J. Catal. 34 (1974), 145.
34. J.F. Le Page, J. Cosyns, P. Courty, E. Freund, J.P. Franck, Y. Jacquin, B. Juguin, C. Marcilly, G. Martino, J. Miquel, R. Montarnal, A. Sugier, H. van Landeghem, in: "Catalyse de Contact", Technip, Paris, 1978, 441.
35. Y. Okamoto, H. Nakamo, J. Shimokawa, S. Teranishi, J. Catal. 50 (1977) 447.
36. S. Harris, R.R. Chianelli, J. Catal. 98 (1986) 17.
37. B.S. Clausen, S. Mørup, H. Topsøe, R. Candia, J. Phys. 37 (1976) 249.
38. H. Topsøe, B.S. Clausen, R. Candia, C. Wivel, S. Mørup, J. Catal. 68 (1981) 433.
39. R.R. Chianelli, M. Daage, Adv. Catal. 40 (1994) 178.
40. M.J. Ledoux, O. Michaux, G. Agostini, P. Panissod, J. Catal. 96 (1985) 189.
41. J.P.R. Vissers, V.H.J. de Beer, R. Prins, J. Chem. Soc., Farad. Trans, 83 (1987) 2145.
42. S.M.A.M. Bouwens, D.C. Koningsberger, V.H.J. de Beer, R. Prins, Bull. Soc. Chim. Belg. 96 (1987) 951.
43. K. Inamura, R. Prins, J. Catal. 147 (1994) 515.
44. H. Topsøe, R. Candia, N.-Y. Topsøe, B.S. Clausen, Bull. Soc. Chim. Belg. 93 (1984) 783.
45. M.W.J. Crajé, PhD thesis, Delft University of Technology, Delft, 1992, ISBN 90-73861-08-X.
46. E.J.M. Hensen, PhD thesis, Eindhoven University of Technology, Eindhoven, 2000, ISBN 90-386-2871-4.

47. C. Wivel, R. Candia, B.S. Clausen, S. Mørup, H. Topsøe, J. Catal. 68 (1981) 453.
48. J.A.R. van Veen, E. Gerkema, A.M. van der Kraan, A. Knoester, J. Chem. Soc. Chem. Commun. (1987) 1684.
49. M.W.J. Crajé, V.H.J. de Beer, J.A.R. van Veen, A.M. van der Kraan, Appl. Catal. A 100 (1993) 97.
50. M.W.J. Crajé, V.H.J. de Beer, A.M. van der Kraan, Appl. Catal. 70 (1991) L7–L13.
51. M.W.J. Crajé, V.H.J. de Beer, A.M. van der Kraan, Hyp. Int. 69 (1991) 795.
52. M.W.J. Crajé, S.P.A. Louwers, V.H.J. de Beer, R. Prins, A.M. van der Kraan, J. Phys. Chem. 96 (1992) 5445.
53. M.W.J. Crajé, V.H.J. de Beer, J.A.R. van Veen, A.M. van der Kraan, J. Catal. 143 (1993) 601.
54. R. Candia, O. Sorensen, J. Villadsen, N.-Y. Topsøe, B.S. Clausen, H. Topsøe, Bull. Soc. Chim. Belg. 93 (1984) 763.
55. H. Topsøe, B.S. Clausen, Catal. Rev. Sci. Eng. 26 (1984) 395.
56. E.J.M. Hensen, V.H.J. de Beer, J.A.R. van Veen, R.A. van Santen, Catal. Lett. 84 (2002) 59.
57. L. Medici, R. Prins, J. Catal. 163 (1996) 38.
58. L. Coulier, V.H.J. de Beer, J.A.R. van Veen, J.W. Niemantsverdriet, J. Catal. 197 (2001) 26.
59. S.M.A.M. Bouwens, F.B.M. van Zon, M.P. van Dijk, A.M. van der Kraan, V.H.J. de Beer, J.A.R. van Veen, D.C. Koningsberger, J. Catal. 146 (1994) 375.
60. H. Shimada, Catal. Today 86, 2003, 17.
61. S.P.A. Louwers, M.W.J. Crajé, A.M. van der Kraan, C. Geantet, R. Prins, J. Catal. 144 (1993) 579.
62. H. Topsøe, B. Hinnemann, J.K. Nørskov, J. V. Lauritsen, F. Besenbacher, P.L. Hansen, G. Hytoft, R.G. Egeberg, K.G. Knudsen, Catal. Today 107-108, 2005, 12.
63. A. Carlsson, M. Brorson, H. Topsøe, J. Catal. 227 (2004) 530.
64. J.V. Lauritsen, S. Helveg, E. Lægsgaard, I. Stensgaard, B.S. Clausen, H. Topsøe, F. Besenbacher, J. Catal. 197 (2001) 1.
65. L.S. Byskov, J.K. Nørskov, B.S. Clausen, H. Topsøe, J. Catal. 187 (1999) 109.
66. S.P.A. Louwers, R. Prins, J. Catal. 139 (1993) 525.
67. H. Shimada, N. Matsubayashi, M. Imamura, T. Sato, Y. Yoshimura, T. Kameoka, K. Masuda, A. Nishijima, Bull. Soc. Chim. Belg. 104 (1995) 353.
68. M. Breysse, J. Bachelier, J.P. Bonnelle, M. Cattenot, D. Cornet, T. Décamp, J.C. Duchet, R. Durand, P. Engelhard, R. Frety, C. Gachet, P. Geneste, J. Grimblot, C. Gueguen, S. Kasztelan, M. Lacroix, J.C. Lavalley, C. Leclercq, C. Moreau, L. De Mourgues, J.L. Olivé, E. Payen, J.L. Portefaix, H. Toulhoat, M. Vrinat, Bull. Soc. Chim. Belg. 96 (1987) 829.
69. H.R. Reinhoudt, C.H.M. Boons, A.D. van Langeveld, J.A.R. van Veen, S.T. Sie, J.A. Moulijn, Appl. Cat. A 207 (2001) 25.
70. B. Scheffer, P.J. Mangnus, J.A. Moulijn, J. Catal. 121 (1990) 18.
71. H.R. Reinhoudt, E. Crezee, A.D. van Langeveld, P.J. Kooyman, J.A.R. van Veen, J.A. Moulijn, J. Catal. 196 (2000) 315.
72. H. Topsøe, Stud. Surf. Sci. Catal. 130 (2000) 1.
73. P.J. Kooyman, J.G. Buglass, H.R. Reinhoudt, A.D. van Langeveld, E.J.M. Hensen, H.W. Zandbergen, J.A.R. van Veen, J. Phys. Chem. B 106 (2002) 11795.

Chapter 2

Experimental techniques

Abstract

A general introduction to and details of the experimental methods (Mössbauer spectroscopy, extended X-ray absorption fine structure (EXAFS), transmission electron microscopy (TEM), and dibenzothiophene HDS activity measurements) are presented. The first-time application of high-pressure ^{57}Co Mössbauer emission spectroscopy and ^{182}W Mössbauer absorption spectroscopy for characterization studies on catalysts is described in detail.

2.1 Mössbauer Spectroscopy

Mössbauer spectroscopy is a technique based on the recoilless emission and resonant absorption of gamma X-rays. The Mössbauer effect was discovered by Rudolph Mössbauer in 1957 [1], and the technique rapidly developed into a powerful tool in solid-state physics and other scientific fields including inorganic chemistry, mineralogy and geochemistry. The technique is used to measure accurately the spacing of nuclear energy levels with a high-energy resolution, better than 1 part in 10^{12}. This enables the use of Mössbauer spectroscopy as a highly sensitive probe of the atomic environment, providing valuable information on oxidation states, magnetic fields, lattice symmetry, and lattice vibrations. Numerous texts describing the fundamentals of Mössbauer spectroscopy [2-6] and reviews on its applications to the study of catalysts [7-10] are available.

An important advantage of Mössbauer spectroscopy is that it makes use of high penetrating power γ-radiation, such that the technique can be applied in an *in situ* manner. Although a typical Mössbauer experiment is relatively inexpensive, it often requires laboratory facilities for preparation and handling of radioactive samples. All ^{57}Co Mössbauer emission spectroscopy (MES) and ^{182}W Mössbauer absorption spectroscopy (MAS) measurements were performed at the Reactor Institute Delft from the Delft University of Technology.

2.1.1 The Mössbauer effect

A stationary and isolated nucleus transits from an excited state to a ground state with the emission of a photon. Due to conservation of momentum the nucleus takes up kinetic recoil energy and the energy of the emitted photon will be lower than the energy required to bring another free nucleus into an excited state. This explains why the Mössbauer effect cannot be observed in free atoms such as in a gas or a liquid. When the atoms of the source and the absorber are embedded in a completely rigid lattice, the recoil momentum is the same as for a free nucleus but is shared by the lattice as a whole. A fraction of the emission and absorption processes occurs without recoil with a probability given in the recoil-free fraction, f:

$$f = e^{-k_\gamma^2 <x^2>} \tag{2.1}$$

where k_γ denotes the wave number of the photon and $<x^2>$ the mean square displacement of the emitting nucleus caused by lattice vibrations. From this equation results that the recoilless fraction will be low in the case of soft vibrational modes, in which $<x^2>$ is relatively large and also for transitions with high k_γ, i.e. for high-energy transitions.

Resonant absorption can only take place if the emitting and the absorbing nuclei are in an identical chemical environment. However, the energy of the emitted photons can be modulated to match the energy levels in the absorber by making use of the Doppler effect. The Doppler effect is the apparent change in the sound or electromagnetic frequencies when there is relative motion between the source and the observer. In the same way, the γ-energy is shifted by giving the source a velocity relative to the absorber. The Mössbauer spectra are recorded by measuring the transmission of the γ-photons through the absorber as a function of the velocity of the source.

2.1.2 ^{57}Co Mössbauer emission spectroscopy

In the present work, MES is used to study the sulfidation of Co in CoMo catalysts. Radioactive ^{57}Co was introduced in the catalysts and was used as a source of the 14.4 keV γ- radiation, as shown in Figure 2.1.

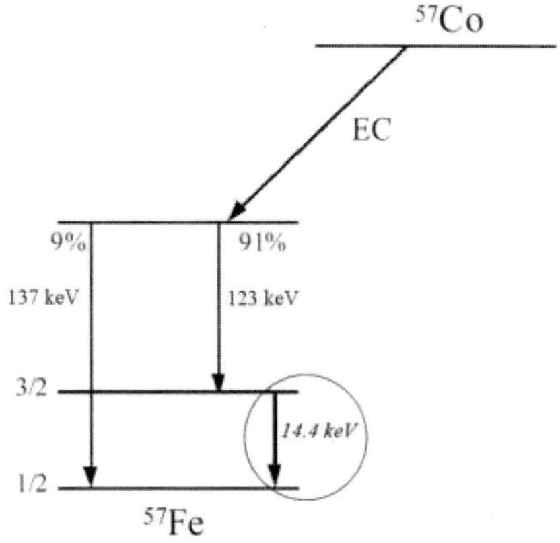

Figure 2.1 *The decay scheme of ^{57}Co.*

The transition from ^{57}Co to ^{57}Fe occurs via electron capture (EC), with 9 % decaying directly to the ground state and 91 % following a two step decay, including the 14.4 keV Mössbauer transition.

The peak positions in a Mössbauer spectrum are sensitive to the interaction of the nucleus with its surroundings, such that different compounds give different spectra. The interactions between the nuclear charge distribution and the extranuclear electric and magnetic fields (hyperfine interactions) are connected to three Mössbauer parameters, i.e. the isomer shift (I.S.), the quadrupole splitting (Q.S.) and the magnetic Zeeman splitting.

The *isomer shift* is a consequence of the electrostatic interaction between the nuclear charge distribution and the s-electrons, which have a finite probability to be found in the region of the nucleus. This interaction results in a slight shift of the nuclear levels of the ground and the excited state, as displayed in Figure 2.2 (a). The shift is in general different for both absorber and source and a Doppler velocity will have to be supplied to the source or absorber to observe resonance. In the MES spectrum, a shift of the peak position with respect to Doppler velocity zero of a reference absorber is observed, providing information on the oxidation state of the atoms of the investigated compound. The same shift of the energy levels can also arise from the thermal motion of the Mössbauer atoms. This phenomenon is named "the second order Doppler shift", but it is usually much smaller than the isomer shift and its variations from compound to compound are very small.

In the 14.4 keV excited state, ^{57}Fe nucleus has a positive quadrupole moment, which reflects the deviation from the spherical charge distribution of the nucleus. The magnitude of the *quadrupole splitting* is proportional to the electric field gradient (EFG), which interacts with the quadrupole moment of the nucleus. The EFG is determined by the chemical environment, i.e. the asymmetrical distribution of electrons of the atom itself and charges of neighboring atoms. Therefore, the splitting shown in the Mössbauer spectrum by two separate peaks, as can be seen in Figure 2.2 (b), provides information about the symmetry around the probe atom.

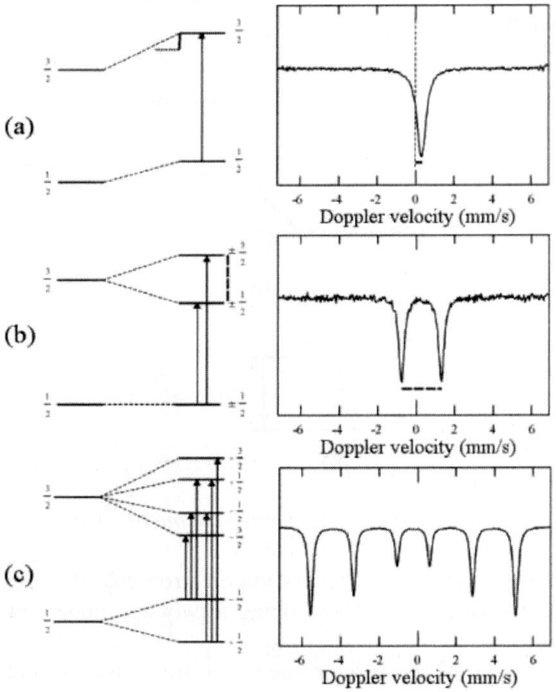

Figure 2.2 *Influence of the hyperfine interactions: (a) isomer shift, (b) quadrupole splitting, and (c) magnetic splitting, on the nuclear energy levels of* ^{57}Fe *and the resulting Mössbauer spectra.*

The *magnetic splitting* arises from the interaction between the nuclear magnetic dipole moment and the local and applied magnetic fields at the nucleus, resulting in the complete removal of the degeneracy of the nuclear energy levels. The ground state (spin 1/2) splits into two and the excited state (spin 3/2) splits into four levels, but only six transitions out of the eight possible are allowed (according to the selection rules: $\Delta m = 0, \pm 1$, where m is the magnetic quantum number). This results in the observation of a symmetric six-line Mössbauer spectrum, as depicted in Figure 2.2 (c), where the isomer shift is given by the center of gravity of the six peaks and the separation between the outer peaks is proportional to the magnitude of the magnetic field.

2.1.3 ^{182}W Mössbauer absorption spectroscopy

Mössbauer spectroscopy utilizing the 100 keV transition of ^{182}W provides a powerful and unique means of studying the structure and bonding of tungsten compounds [11]. The transition is from an excited nuclear state with spin I = 2 to a I = 0 ground state, the excited state being fed by the decay of ^{182}Ta (half-life 115 days), as shown in Figure 2.3.

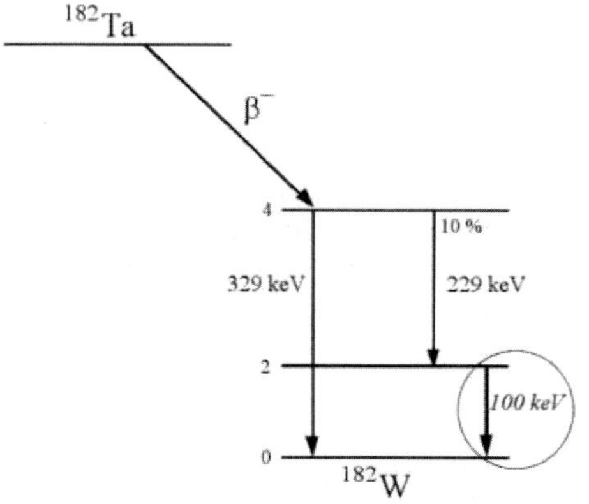

***Figure 2.3** Simplified decay scheme of ^{182}Ta.*

The quadrupole moment of the excited state is large [12] and this enables the accurate measurement of the quadrupole splitting. The presence of an axially symmetric EFG results in a hyperfine pattern consisting of three lines ($I_z = \pm 2, \pm 1, 0$), while for an axially non-symmetric EFG the degeneracy of the sublevels is lifted and a five-line spectrum is obtained. The electric quadrupole energies of the five excited states for I = 2 are given by the following equations [13,14]:

The eQV_{ZZ} term denotes the quadrupole interaction constant, where Q is the quadrupole moment of the nucleus, V_{ZZ} is the z component of the EFG, and e is the electronic charge. In the case of axially non-symmetric EFG, the quadrupole splitting depends on the asymmetry parameter (η).

The isomer shift depends on a nuclear factor $\delta R = R_e - R_g$, where R is the radius of the nucleus and the subscripts *e* and *g* refer to the excited and ground nuclear states respectively. For a given nucleus, δR is a constant and the isomer shift is directly proportional to the s-electron density at the nucleus. For the 100 keV transition of ^{182}W, the value of δR is extremely small and the observation of shifts greater than the experimental errors is not expected [14].

$$E(+2) = \frac{1}{4} \cdot eQV_{ZZ} \tag{2.2}$$

$$E(+1) = -\frac{1}{8} \cdot eQV_{ZZ} \cdot (1+\eta) \tag{2.3}$$

$$E(0) = -\frac{1}{4} \cdot eQV_{ZZ} \cdot \sqrt{\left(1 + \frac{\eta^2}{3}\right)} \tag{2.4}$$

$$E(-1) = -\frac{1}{8} \cdot eQV_{ZZ} \cdot (1 - \eta) \tag{2.5}$$

$$E(-2) = \frac{1}{4} \cdot eQV_{ZZ} \cdot \sqrt{\left(1 + \frac{\eta^2}{3}\right)} \tag{2.6}$$

2.1.4 Experimental details

In the present work, Mössbauer spectroscopy is used in two different ways: Mössbauer emission spectroscopy in which a single line absorber ($K_4Fe(CN)_6{\cdot}3H_2O$) is moved to investigate an unknown source containing ^{57}Co and Mössbauer absorption spectroscopy, where a single line ^{182}Ta source is moved to scan the unknown energy levels in a W containing absorber. In Figure 2.4 the two different spectroscopic modes are drawn schematically.

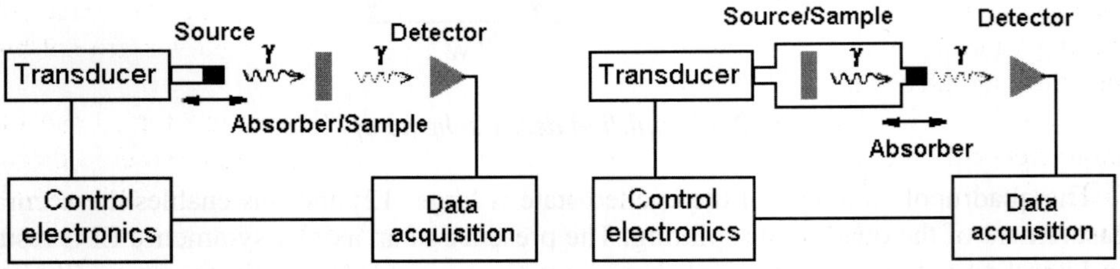

Figure 2.4 *Schematic view of the experimental setup for Mössbauer absorption (left) and emission (right) studies.*

Because of the large penetration power of the γ-rays, both MES and MAS techniques can be used to study catalysts under *in situ* conditions. For the MES measurements the samples were sulfided in a flow of 60 cm^3 min^{-1} of 10% H_2S/H_2 mixture at pressures up to 4 MPa and temperatures up to 773 K in a high-pressure Mössbauer *in situ* reactor, similar to the reactor described in detail in [15]. This reactor offers the possibility of studying the catalysts under conditions which mimic those encountered in industrial practice. A state-of-the-art high-pressure *in situ* cell was developed and manufactured at the Reactor Institute Delft for the ^{182}W MAS measurements. Due to the high energy of the ^{182}W Mössbauer transition, cooling of the source and the absorber is required to obtain a measurable resonant absorption. The *in situ* cell allows the sulfidation of the samples under the same high pressure and high temperature conditions stated above and the consequent MAS measurements at cryogenic temperatures down to 4.2 K.

The ^{57}Co MES spectra were recorded at room temperature and at the treatment pressure, using a constant acceleration spectrometer in a triangular mode with a moving single-line $K_4Fe(CN)_6{\cdot}3H_2O$ absorber enriched in ^{57}Fe. The velocity scale was calibrated with a ^{57}Co:Rh source and a sodium nitroprusside (SNP) absorber. Zero velocity corresponds to the peak position of the $K_4Fe(CN)_6{\cdot}3H_2O$ absorber measured with the ^{57}Co:Rh source; positive velocities correspond to the absorber moving toward the source. The spectra were analyzed with a Lorentzian fitting procedure as described in [16].

The ^{182}W MAS spectra were recorded at liquid helium temperature using a ^{182}Ta in Ta metal source which was moved in a sinusoidal mode. The source was prepared at the Reactor Institute Delft by irradiating metallic Ta with thermal neutrons. The transmitted radiation was recorded with a high-purity Ge detector and stored in a 1024 channel analyzer. A metallic Fe foil was used for velocity calibration with the ^{57}Co:Rh source. The spectra were analyzed by least-squares fits with a proper set of Lorentzian-shape lines.

The Mössbauer spectra encountered in practice are subject to statistical noise. The goodness of the fits was estimated from the chi-square quantity using *MossWinn 3.0i* software [35]:

$$\chi^2(v) = \sum_{i=1}^{h} \frac{(W_i - f_i(v))^2}{W_i} \tag{2.7}$$

where Wi denotes the counts in the ith channel of the Mössbauer spectrum, h is the number of channels, and fi is the value of the fitting function (e.g. sum of Lorentzians subtracted from the base line) corresponding to the ith channel of the spectrum. The parameter χ^2 was used to compare the goodness of the fit of a spectrum fitted with several different sets of starting values. Typically, the resulted differences in Mössbauer parameters are not higher than the estimated standard deviations.

MossWinn 3.0i was employed to determine the standard deviation for all the fitting parameters by using Monte Carlo iterations. For the ^{57}Co MES spectra a standard deviation (SD) of 0.03 mm s^{-1} was calculated for the isomer shift and quadrupole splitting values, whereas an SD of 0.05 mm s^{-1} was determined for the linewidth. The absolute SD of the spectral contribution (A) values was about 5%. No significant differences in the SD values estimated for samples having different Co loadings were observed.

For the ^{182}W MAS spectra an SD of 0.2 mm s^{-1} was determined for the Q.S. parameters and an SD value of 0.4 mm s^{-1} was obtained for the linewidths. The absolute SD of the spectral contribution (A) values was about 15%.

2.2 Extended X-ray Absorption Fine Structure (EXAFS)

2.2.1 Introduction to EXAFS

The fine structure in the X-ray absorption coefficient above an absorption edge of an element allows determination of structural parameters of the local atomic surroundings. The interference effects observed in the X-ray absorption spectrum give information on the number and type of neighbors as well as on the interatomic distances and structural disorder [17,18].

The photoelectric effect dominates the attenuation process in the X-ray regime where EXAFS occurs. In the photoelectric absorption part of the photon energy is used to overcome the binding energy of the electron and the rest is given to the electron as kinetic energy. When a photon beam is transmitted through a specific sample, its incident intensity will be decreased according to Lambert's law:

$$I_t = I_0 \cdot e^{-\mu(E) \cdot x} \tag{2.8}$$

where I_0 and I_t are the intensities of the incoming and transmitted beam, respectively, μ is the absorption coefficient, and *x*, the sample thickness.

The most common way of performing an X-ray absorption experiment is to record the incident and transmitted beam as a function of photon energy. A smooth absorption coefficient, monotonically decreasing with the energy of the photon, is observed before the edge, while at an energy equal to the threshold energy of a bound electron a discrete rise in the absorption will be measured due to the ejection of the bound electron. For photon energies larger than the binding energy, the ejected photoelectron travels as an outgoing spherical wave, with the photoelectron wave number *k*, equal to:

$$k = \sqrt{\frac{2m}{\hbar^2}(E - E_b)} \tag{2.9}$$

where E represents the photon energy, E_b is the binding energy of the electron, and *m* is the mass of the electron.

For isolated atoms, the photoelectron is allowed to travel unhampered and μ decreases as a function of energy beyond the edge, as shown in Figure 2.5 (a). If neighbor atoms are present in the surroundings of the absorbing atom, the outgoing photoelectron wave will be scattered by these atoms and the final state of the electron will be determined by the summation of the outgoing wave and the backscattered electron waves. The interference pattern can be constructive or destructive, depending upon the kinetic energy of the incoming photon, resulting in the modulation of the absorption spectrum after the edge, as illustrated in Figure 2.5 (b).

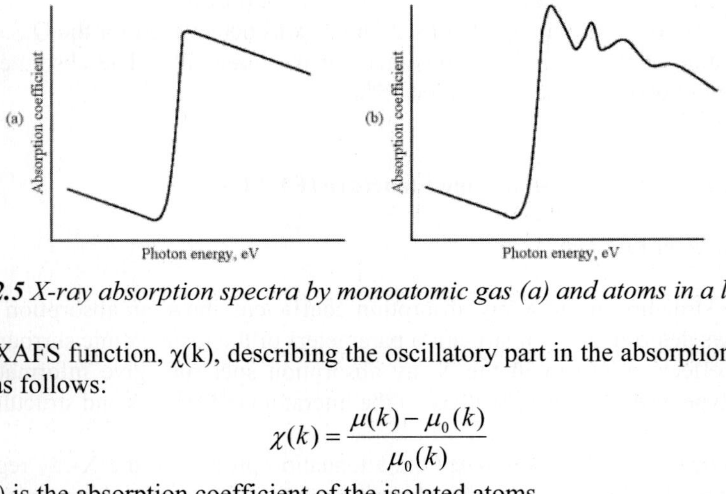

Figure 2.5 *X-ray absorption spectra by monoatomic gas (a) and atoms in a lattice (b).*

The EXAFS function, χ(k), describing the oscillatory part in the absorption coefficient, is defined as follows:

$$\chi(k) = \frac{\mu(k) - \mu_0(k)}{\mu_0(k)} \tag{2.10}$$

where $\mu_0(k)$ is the absorption coefficient of the isolated atoms.

In the "small atom approximation", in which the atomic diameter is small compared to the interatomic distance, the outgoing electron is treated as a plane wave. The relation between χ(k) and the structural parameters is obtained using the atomic potentials from the muffin-tin approximation and taking into account only single scattering [19,20]:

$$\chi(k) = \sum_j \frac{N_j \cdot F_j(k)}{k \cdot R_j^2} \cdot G_j(k) \cdot \sin(2k \cdot R_j + \phi_j(k)) \qquad (2.11)$$

The first part of the EXAFS function contains information on the amplitude of each scattering contribution, while the term "$\sin(2k \cdot R_j + \phi_j(k))$" contains frequency information, where $\Phi_j(k)$ is the total phase shift experienced by the photoelectron. The amplitude is proportional to the number of atoms N_j in the coordination shell and inversely proportional to the square of the coordination distance R_j. F_j is the backscattering amplitude from each neighboring atom in the jth shell and $G_j(k)$ is a term containing corrections for the finite electron mean free path length and disorder:

$$G_j(k) = S_0^2(k) \cdot \exp(-2k^2 \cdot \sigma_j^2) \cdot \exp(\frac{-2R_j}{\lambda}) \qquad (2.12)$$

$S_0(k)$ is the correction for relaxation effects in the emitting atom, while the term $\exp(-2k^2 \cdot \sigma_j^2)$ accounts for dynamic disorder (caused by lattice vibrations) and static disorder if atoms of the same coordination shell have slightly different distances to the central atom. The term σ_j^2 represents the mean-squared displacements of atoms in the sample. To account for the inelastic losses in the scattering process of the electron when it travels through the solid, an additional term – $\exp(-2R_j / \lambda)$ – is added, in which λ is the inelastic mean free path of the electron.

When both the backscattering amplitude and the phase shift are known, the structural parameters N_j, R_j and σ_j can be derived from χ(k). Each element has specific phase function and backscattering amplitude that can be calculated [21] or obtained by measuring well-known reference compounds.

2.2.2 EXAFS measurements and data analysis

The EXAFS spectra, measured at the Co K-edge (7.709 keV) and the Mo K-edge (20 keV), were obtained at the Dutch-Belgian Beamline (DUBBLE) at the European Synchrotron Radiation Facility (ESRF), Grenoble, France. The electron energy and ring current were 6 GeV and 150-200 mA, respectively. The catalysts were stepwise sulfided in stainless-steel tubular reactors that were subsequently flushed with Ar and opened in a glove box. Self-supporting wafers of the sulfided catalysts were pressed in the glove box and brought in an environmental cell. The thickness of the wafer was chosen to give an absorbance (μx) of about 2.5 in the Mo K-edge region to ensure an optimal signal-to-noise ratio. Because of the low Co concentration in the catalyst, the absorbance for the Co K-edge measurements was about 3. Three scans of each sample were recorded in transmission mode.

The main experimental challenges for a good measurement of μ(k) are getting an X-ray source that can be tuned in energy, and high-quality detectors of X-ray intensity. A synchrotron is used as a source, which provides a full range of x-ray wavelengths, and a Si (111) channel-cut monochromator that uses Bragg diffraction to select a particular energy. The experimental setup for an EXAFS measurement is shown in Figure 2.6. Most of the spectra were recorded in the transmission mode, while fluorescence geometry was used with samples having low concentrations of the studied element.

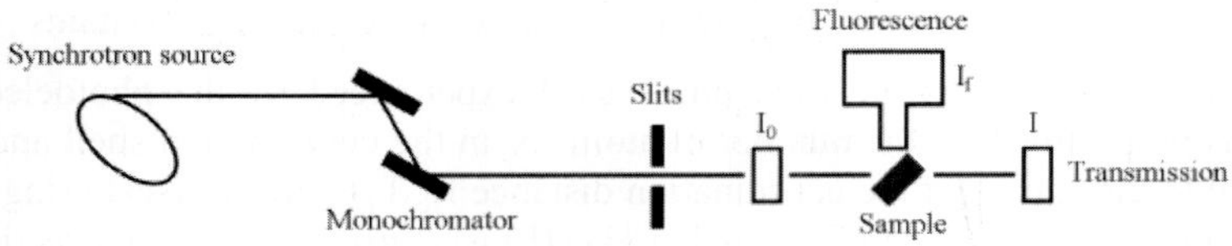

Figure 2.6 *Schematic view of the experimental setup for EXAFS studies. I_0 is the ionization chamber detecting the incident beam, I is the ionization chamber detecting the transmitted beam, and I_f the fluorescence ionization chamber.*

The typical EXAFS spectrum can be divided in four regions, including about 200 eV of pre-edge, the edge itself, the XANES (X-ray Absorption Near Edge Structure) region that is about 50 eV wide, and the EXAFS region (500-2000 eV wide), as indicated in Figure 2.7.

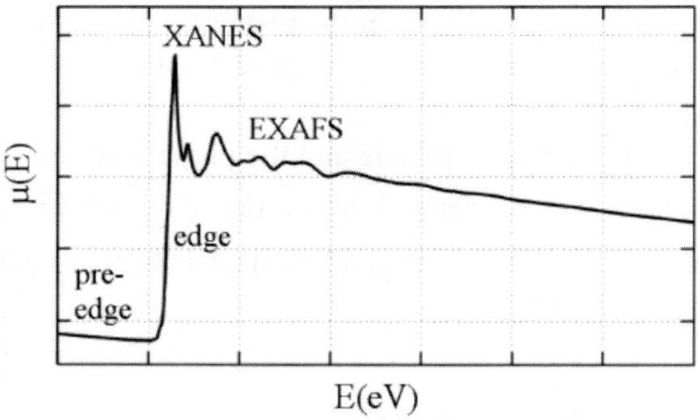

Figure 2.7 *Typical X-ray absorption spectrum: μ(E) versus energy relative to the absorption edge.*

All the above-mentioned EXAFS regions have different requirements concerning the number of points and the signal-to-noise ratio. A large step size was used in the pre-edge region, while many points (small step size) were recorded at the edge, which is used to calibrate the energy scale. The XANES and EXAFS regions were also measured with a relatively small step size to ensure sufficient signal-to-noise ratio.

In order to extract the structural information contained in the measured spectra, the χ(k) function has to be isolated by data reduction techniques and the phase and the backscattering functions have to be found. The XDAP program – version 2.2.2 [22,23] – was used for data manipulation and data analysis.

The EXAFS function in k-space was obtained from the X-ray absorption spectra by subtracting a Victoreen curve that was fitted to the pre-edge region, followed by a cubic spline background removal [24]. Normalization was performed through division by the height of the edge jump, which is proportional to the number of absorbing atoms. The obtained $\chi(k)$ function results in a radial distribution (FT function) after Fourier transformation. The transformations were executed over the largest possible k-range whose limits were chosen in nodes of the EXAFS function to minimize cutoff effects.

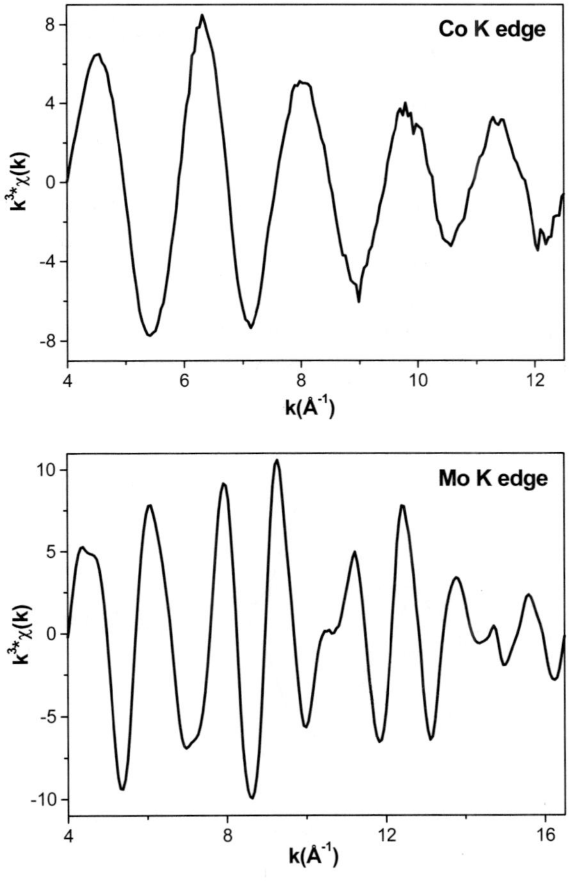

Figure 2.8 *Co K-edge and Mo K-edge raw EXAFS data (k^3-weighted) of sulfided $CoMo/Al_2O_3$ catalyst.*

In the fitting procedure phase shifts and backscattering amplitudes from reference compounds were used [25,26]. For the Co-S EXAFS signal CoS_2 was employed and for the Co-Co contribution the Ni-Ni coordination in NiO was chosen. The Mo-S and Mo-Mo first shell coordinations of MoS_2 were used as reference for Mo-S and Mo-Mo EXAFS functions. Finally, for the Co-Mo and Mo-Co signals the Ni-Mo coordination in $((C_6H_5)_4P)_2Ni(MoS_4)_2$ was selected. The use of a Ni absorber instead of Co is justified since phase shifts and backscattering amplitudes of neighboring elements such as Co and Ni hardly differ [27].

The EXAFS results were produced by fitting in R-space until a satisfactory fit was obtained for k^1- and k^3-weighted spectra in R-space as well as k-space. In addition, the similarity in R-space had to be adequate for both the absolute value and the imaginary part of the function. The various FT ranges used were chosen between 4 and 12.5 Å^{-1} for Co K-edge and between 4 and 16.5 Å^{-1} for Mo K-edge (Figure 2.8).

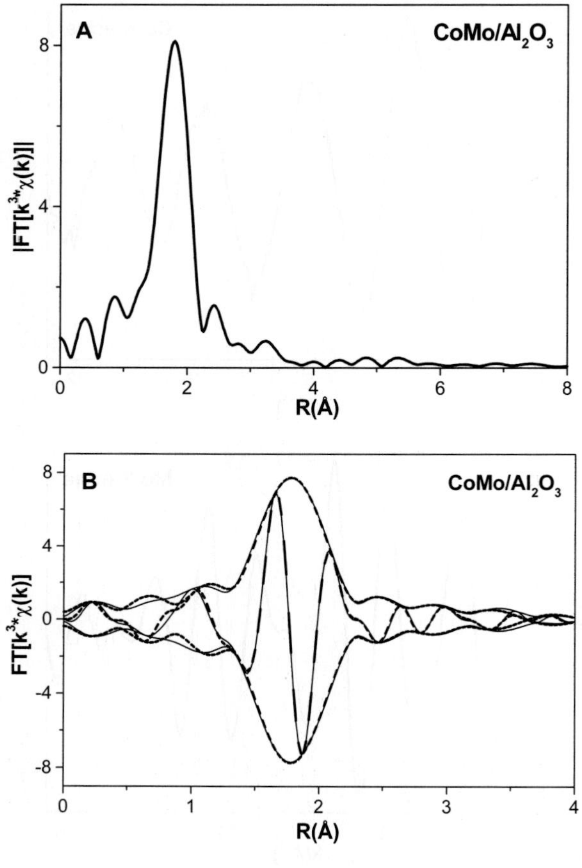

Figure 2.9 *Co K-edge: (A) Absolute part of k^3-weighted Fourier transform of sulfided $CoMo/Al_2O_3$; (B) k^3-weighted FT-function (solid line) and best fit (dotted line).*

In Figure 2.9 the absolute part of Fourier transformed (k^3-weighted) Co K-edge EXAFS function of a calcined $CoMo/Al_2O_3$ catalyst sulfided at 673 K under high-pressure conditions is presented. The k^3-weighted FT-function (solid line) together with the best fit (dotted line) are also shown. The data obtained at the Mo K-edge with the same sample, treated under the same conditions, are presented in Figure 2.10. As can be observed in Figures 2.8, 2.9 and 2.10 the quality of the data at both edges and of the computer fits was high. The quality of the fits was estimated from the value of the goodness of the fit (ε_v^2) as defined in [33].

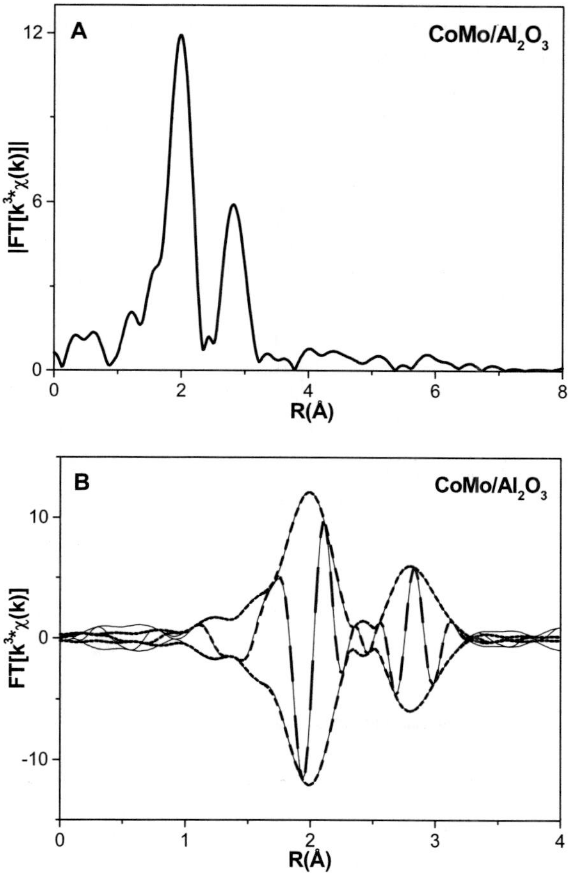

Figure 2.10 *Mo K-edge: (A) Absolute part of k^3-weighted Fourier transform of sulfided $CoMo/Al_2O_3$; (B) k^3-weighted FT-function (solid line) and best fit (dotted line).*

In order to check the suitability of the four-shell fit applied to the Co K-edge EXAFS function (Figure 2.9), a three-shell fit (including the Co-S, Co-Co(1), and Co-Mo contributions) is presented in Figure 2.11 A. The k^3-weighted FT-function of the difference file (EXAFS function minus best Co-S, Co-Co(1), and Co-Mo contributions) is shown in Figure 2.11 B (solid line). A peak at around 3.2 Å (not phase corrected) is observed in the difference file together with some remnant of the background signal below 1 Å). The best Co-Co(2) contribution is also shown in Figure 2.11 B (dotted line), the one-shell fit fairly coinciding with the peak ascribed to the Co-Co(2) coordination in the four-shell fit. The result suggests that the inclusion of a Co-Co(2) shell in the fit is appropriate.

The number of free parameters that may be optimized can be calculated from the Nyquist theorem [34]:

$$\text{Number of free parameters} = \frac{2\Delta k \Delta R}{\pi} + 1 \tag{2.13}$$

The maximum number of independent parameters that can be determined from the present set of Co K-edge EXAFS spectra is 22.6. In the four-shell fit, 16 independent parameters were used, which is safely below the indicated limit.

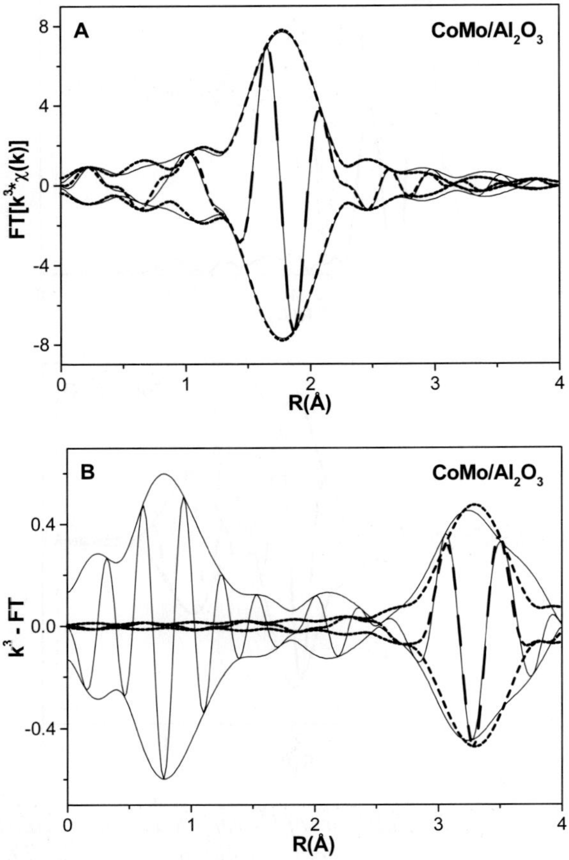

Figure 2.11 *Co K-edge: (A) k^3-weighted FT-function (solid line) and best three-shell fit (dotted line); (B) EXAFS data minus best Co-S, Co-Co(1), and Co-Mo contributions (solid line) and best fit Co-Co(2) contribution (dotted line).*

To prove the appropriateness of the addition of a Co-Mo contribution in the fit, a two-shell fit with only Co-S and Co-Co(1) contributions was also attempted (Figure 2.12, A). The k^3-weighted FT-function of the difference file (EXAFS function minus best Co-S and Co-Co(1) contributions) is shown in Figure 2.12 B (solid line). A sizable peak at around 2.6 Å (not phase corrected) is present. The Co-Mo contribution fit is also shown in Figure 2.12 B (dotted line); the fit takes into account the peak assigned to the Co-Mo shell in the four-shell fit. It is concluded that the Co-Mo contribution cannot be excluded from the fit of these EXAFS spectra.

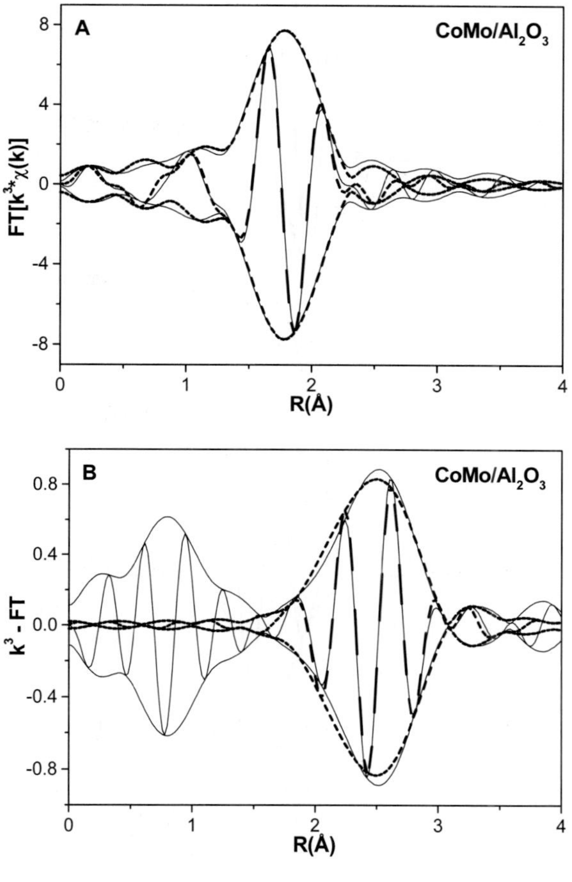

Figure 2.12 *Co K-edge: (A) k^3-weighted FT-function (solid line) and best two-shell (Co-S, Co-Co(1)) fit (dotted line); (B) EXAFS data minus best Co-S and Co-Co(1) contributions (solid line) and best fit Co-Mo contribution (dotted line).*

From the investigation of the two-shell fit including the Co-S and Co-Mo contributions (Figure 2.13, A), the addition of a Co-Co(1) shell was evaluated. The k^3-weighted FT-function of the difference file (EXAFS function minus best Co-S and Co-Mo contributions) is shown in Figure 2.13 B (solid line). The fairly good correspondence of the best Co-Co(1) fit (Figure 2.13 B - dotted line) to the k^3-weighted FT of the difference file suggests the presence of a Co-Co(1) contribution. Also from the clear presence of a Co-Co(2) contribution and from the systematic trends observed in the EXAFS results, the inclusion of the Co-Co(1) shell in the fit was deemed necessary.

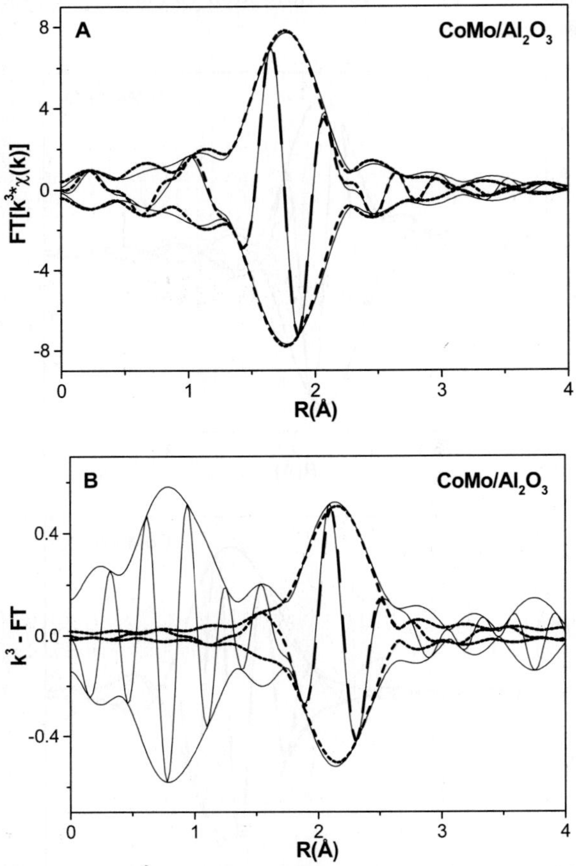

Figure 2.13 *Co K-edge: (A) k^3-weighted FT-function (solid line) and best two-shell (Co-S, Co-Mo) fit (dotted line); (B) EXAFS data minus best Co-S and Co-Mo contributions (solid line) and best fit Co-Co(1) contribution (dotted line).*

In a similar manner, the correctness of including a Mo-Co contribution to the Mo K-edge EXAFS function (Figure 2.10) was evaluated by comparing a three-shell fit (Mo-S, Mo-Mo and Mo-Co contributions) and a two-shell fit (Mo-S and Mo-Mo contributions). The two fits and k^3-weighted FT-function of the difference file (EXAFS function minus best Mo-S and Mo-Mo contributions) are shown in Figure 2.14 A and Figure 2.14 B, respectively. A peak around 2.2 Å (not phase corrected) is present for which inclusion of the best Mo-Co contribution gives a reasonable description of the missing contribution.

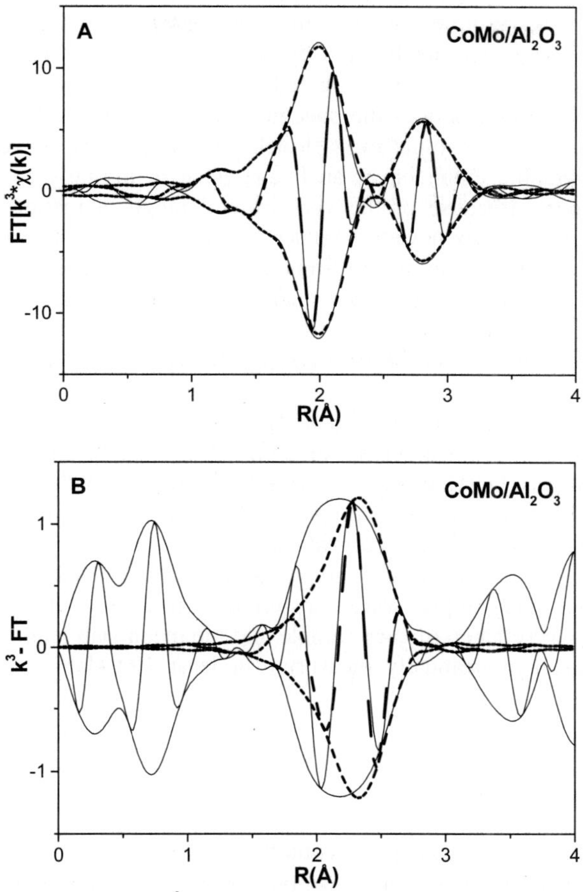

Figure 2.14 *Mo K-edge: (A) k^3-weighted FT-function (solid line) and best two-shell fit (dotted line); (B) EXAFS data minus best Mo-S and Mo-Mo contributions (solid line) and best fit Mo-Co contribution (dotted line).*

Also in this case, a three-shell fit with 12 independent parameters is within the limit determined from the Nyquist theorem for the Mo K-edge EXAFS spectra (32.8).

2.3 Transmission electron microscopy

Transmission electron microscopy (TEM) is a very powerful imaging technique for obtaining information on particle size and shape of microstructures. In TEM, transmitted and diffracted electrons are used to characterize the internal structure of a wide variety of materials [28]. Because the wavelength of electrons is about 100.000 times shorter than the wavelength of visible light, electron microscopy has a much greater resolving power than light microscopy [29].

Similar to a conventional light microscope, a transmission electron microscope consists of an illumination system, sample stage and imaging system. High-energy electrons (100-400 keV) from an electron gun are focused by condensor lenses to produce parallel rays that illuminate the specimen to be investigated. Electron optics is used to magnify the electron intensity distribution behind the specimen, resulting in the formation of a two-dimensional projection of the sample mass. Finally, the image can be recorded by direct exposure of a photographic emulsion or an image plate or digitally by a CCD (charge-coupled device) camera. Because of the very limited range in matter of electrons, the samples for electron microscopy have to be placed in vacuum inside the instrument and are required to be very thin, usually in form of films mounted on fine-meshed grids.

By making use of the image mode, the microstructure of the particles can be studied, while the crystalline structure is studied by the diffraction mode. In addition, the chemical composition of small volumes can be obtained by detection of X-rays emitted from the film.

TEM measurements were performed at the National Center for High Resolution Electron Microscopy at Delft University of Technology, Delft. The micrographs were obtained using a Philips CM30T electron microscope equipped with a field emission gun operated at 300 keV [30]. Samples were prepared by mounting a few drops of a suspension of ground catalyst in *n*-hexane on a microgrid carbon polymer supported on a Au grid (400 mesh). For the TEM measurements, the samples sulfided for EXAFS measurements were used. After preparation, the samples were transferred to glass ampoules in a glove box. At least ten micrographs were taken of each sample and the mean slab length and the stacking degree were determined by manually measuring at least 300 slabs per sample in three representative images.

2.4 HDS activity measurements

HDS of model reactants is preferentially employed to evaluate the catalytic performance since the aromatics and the N-containing molecules in the real feedstock could potentially give rise to complications related to the poisoning of the active sites. The molecular size and the structure of the sulfur-containing compounds have critical influence on the hydrodesulfurization reactivity, dibenzothiophene (DBT) being more difficult to desulfurize than benzothiophene and thiophene [31].

Medium-pressure gas-phase DBT HDS measurements were performed in a stainless-steel reactor. DBT dissolved in *n*-decane (1.0 wt.% DBT) was fed by an HPLC pump at a rate of 0.1 cm^3 min^{-1} into a hydrogen stream of 0.5 dm^3 min^{-1} (final DBT concentration 200 ppm [32]). The reactor was loaded with about 20 mg of catalyst, diluted with 5 g SiC to obtain a catalyst bed height of about 3 cm. Prior to activity measurements, the catalyst was activated at the desired pressure in a flow of 60 cm^3 min^{-1} of a mixture of 10 vol.% H_2S in H_2, whilst heating to 673 K at a rate of 6 K min^{-1} followed by an isothermal period of 1 h. Subsequently, the reactant feed was passed over the catalyst at a total pressure of 3 MPa. Typically, 8 h were allowed at each reaction condition for the catalyst to stabilize. Gaseous products were analyzed by on-line gas chromatography (HP 5890, CP-Sil-5CB, FID). For the calculation of the rate constant (k_{HDS}) first-order kinetics in DBT were assumed, according to the formula:

$$k_{HDS} = -\frac{F}{W} \cdot \ln(1 - X) \qquad (2.14)$$

in which F [mol s^{-1}] is the feed gas flow, W [kg] is the amount of Mo in the catalyst, and X the conversion of the reactant. The apparent activation energies (E_{act}) and the pre-exponential factors (ν_{pre}) were evaluated from a plot of the reaction rate as a function of reaction temperature. Three data points (533, 553 and 573 K) were collected to estimate the kinetic parameters. The typical Arrhenius plot obtained with a $CoMo/Al_2O_3$ catalyst is presented in Figure 2.15 (the error bars depict the standard deviation of the plotted points). The standard deviation of the activation energies values varied between 2 kJ/mol and 7 kJ/mol, a typical value of the SD was 3 kJ/mol.

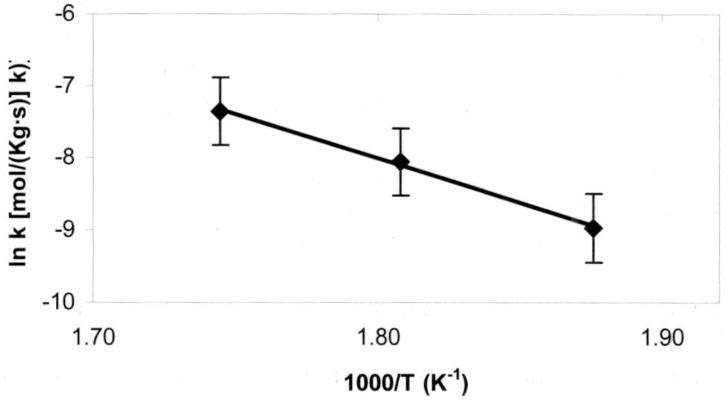

Figure 2.15 *Arrhenius plot (error bars indicate standard deviation in ln k) for a $CoMo/Al_2O_3$ catalyst over the 533-573 K temperature interval.*

References

1. R.L. Mössbauer, Z. Physik 151 (1958) 124.
2. G.K. Wertheim, Mössbauer effect: Principle and Applications, Academic Press, New York, 1964.
3. V.I. Goldanskii, R.H. Herber, Chemical Applications of Mössbauer Spectroscopy, Academic Press, New York, 1968.
4. N.N. Greenwood, T.C. Gibb, Mössbauer Spectroscopy, Chapman and Hall, London, 1971.
5. G.J. Long, J.G. Stevens, Industrial Applications of the Mössbauer Spectroscopy, Plenum Press, New York, 1985.
6. T.E. Cranshaw, B.W. Dale, G.O. Longworth, G.E. Johnson, Mössbauer Spectroscopy and its Applications, Cambridge University Press, Cambridge, 1985.
7. J.A. Dumesic, H. Topsøe, Adv. Catal. 26 (1977) 121.
8. A.M. van der Kraan, J.W. Niemantsverdriet, in Industrial Applications of the Mössbauer Effect, G.J. Long and J.G. Stevens (Eds.), Plenum Press, New York, 1985, p. 609.

9. F.J. Berry, in Spectroscopic Characterization of Heterogeneous Catalysts, J.L.G. Fierro (Ed.), Elsevier, Amsterdam, 1990, p. A 299.
10. J.W. Niemantsverdriet, W.N. Delgass, Topics in Catal. 8 (1999) 133.
11. G.M. Bancroft, R.E.B. Garrod, A.G. Maddock, Inorg. Nucl. Chem. Letters 7 (1971) 1157.
12. O. Hansen, M.C. Olsen, O. Skilbreid, B. Elbek, Nucl. Phys. 25 (1961) 634.
13. N. Sikazono, H. Takekoshi, T. Shoji, J. Phys. Soc. Japan 20 (1965) 271.
14. A.G. Maddock, R.H. Platt, A.F. Williams, R. Gancedo, J. Chem. Soc., Dalton Trans. 12 (1974) 1314.
15. M.W.J. Crajé, A.M. van der Kraan, J. van de Loosdrecht, P.J. van Berge, Catal. Today 71 (2002) 369.
16. M.W.J. Crajé, PhD thesis, Delft University of Technology, Delft, 1992, ISBN 90-73861-08-X.
17. D.C. Koningsberger, R. Prins (Eds.), X-ray Absorption, Wiley, New York, 1987.
18. B.K. Teo, EXAFS: Basis principles and data analysis, Springer-Verlag, Berlin, 1986.
19. E.A. Stern, Phys. Rev. B 10 (1974) 3027.
20. C.A. Ashley, S. Doniach, Phys. Rev. B 11 (1975) 1279.
21. J.M. de Leon, J.J. Rehr, S.I. Zabinsky, R.C. Albers, Phys. Rev. B 44(9) (1991) 4146.
22. J.B.A.D van Zon, D.C. Koningsberger, H.F.J. van 't Blik, D.E. Sayers, J. Chem. Phys. 82 (1985) 5742.
23. P.S. Kirlin, F.B.M. van Zon, D.C. Koningsberger, B.C. Gates, J. Phys. Chem. 94 (1990) 8439.
24. J.W. Cook, D.E. Sayers, J. Appl. Phys. 52 (1981) 5024.
25. M.W.J. Crajé, S.P.A. Louwers, V.H.J. de Beer, R. Prins, A.M. van der Kraan, J. Phys. Chem. 96 (1992) 5445.
26. S.M.A.M. Bouwens, R. Prins, V.H.J. de Beer, D.C. Koningsberger, J. Phys. Chem. 94 (1990) 3711.
27. B.K. Teo, P.A. Lee, J. Am. Chem. Soc. 101 (1979) 2815.
28. D. B. Williams, C. B. Carter, Transmission Electron Microscopy: a Textbook for Materials Science, Plenum Press, New York, 1996.
29. J.W. Niemantsverdriet, Spectroscopy in Catalysis: an Introduction, VCH-Verlag, Weinheim, 2000.
30. P.J. Kooyman, J.G. Buglass, H.R. Reinhoudt, A.D. van Langeveld, E.J.M. Hensen, H.W. Zandbergen, J.A.R. van Veen, J. Phys. Chem. B 106 (2002) 11795.
31. H. Topsøe, B.S. Clausen, F.E. Massoth, in: J.R. Anderson, M. Boudart (Eds.), Hydrotreating Catalysts, in: Catal. Science Technol., vol. 11, Springer, Berlin, 1996.
32. W.R.A.M. Robinson, J.A.R. van Veen, V.H.J. de Beer, R.A. van Santen, Fuel Proc. Technol. 61 (1999) 89.
33. F.W. Lytle, D.E. Sayers, E.A. Stern, Physica B 158 (1988) 701.
34. E.A. Stern, Phys. Rev. B 48 (1993) 9825.
35. Z. Klencsár, Nucl. Instr. Meth. B 129 (1997) 527.

Chapter 3

High-pressure sulfidation of calcined $CoMo/Al_2O_3$ hydrodesulfurization catalysts

Abstract

The influence of the sulfiding pressure on the structure and activity of calcined $CoMo/Al_2O_3$ catalysts was studied by Mössbauer emission spectroscopy (MES), extended X-ray absorption fine structure (EXAFS), transmission electron microscopy (TEM) and dibenzothiophene hydrodesulfurization (HDS) activity measurements. Sulfidation at elevated pressure (4 MPa) leads to a much higher HDS activity than upon 0.1 MPa sulfidation. Similarly, the HDS activity increases when after 0.1 MPa sulfidation (673 K) the sulfidation pressure is increased to 4 MPa. The average slab size (~2.8 nm) and stacking degree (~1.4) do not depend on the sulfidation pressure. EXAFS data point to a higher rate of Co and Mo sulfidation at elevated pressure. Although this leads to a somewhat more aggregated form of Co-sulfide particles at intermediate temperatures compared to the case of 0.1 MPa sulfidation, redispersion takes place to small Co-sulfide species on the MoS_2 edges. The spectroscopic data of such stepwise sulfided series support the supposition that sulfidation at 4 MPa leads to a Type II CoMoS phase whereas 0.1 MPa sulfidation results in a less active Type I phase.

3.1 Introduction

The development of improved hydrodesulfurization (HDS) catalysts is essential in order to meet the new environmental legislation requirement for cleaner fuels, having a maximum sulfur content of 10 ppm [1]. Mixed CoMo and NiMo sulfides supported on Al_2O_3 are the most important HDS catalysts. Major progress in our understanding of the synergy between Co and Mo has been the development of the Co-Mo-S model. Amongst others, *in situ* Mössbauer Emission Spectroscopy (MES) has been instrumental to develop a structural model for CoMo sulfide catalysts, i.e. the Co-Mo-S phase, in which Co atoms are located at the edges of MoS_2 slabs [2-4]. Recently, we performed the very first high-pressure *in situ* MES study of the sulfidation of CoMo catalysts [5,6], confirming the applicability of the Co-Mo-S structural model under industrial-like conditions.

The initial Co–Mo–S model was refined by Crajé et al. [7-9], after identical MES spectra observed in Co and CoMo catalysts indicated that no simple relation exists between the amount of Co-Mo-S and the activity [10,11]. Combined MES and extended X-ray absorption fine structure (EXAFS) studies showed that Co-sulfide species at the MoS2 edges differ in particle size and/or ordering. Co-sulfide species in the Co–Mo–S phase exhibit a high quadrupole splitting (Q.S.) value, while smaller Q.S. values point to the presence of larger Co sulfide species [12].

Because important changes in the catalyst structure may occur as a function of the reaction conditions [13,14], it is important to characterize the structure of the catalyst under temperature and pressure conditions relevant to industrial practice [15]. In a systematic manner, we follow the formation and evolution of the active phase as a function of temperature and pressure. Although the MES spectra of the Co-sulfide species in Co and CoMo catalysts can be indistinguishable, we should be able to identify a sulfidation pattern that characteristically occurs when the Co-Mo-S structures are formed. In the same time we examine the sulfidability of the Co atoms under conditions closer to industrial practice than those used in earlier studies performed under atmospheric pressure conditions [12]. Our previous results obtained with an identical catalyst sulfided directly at high-pressure [5,6] have shown that activation at 673 K and 4 MPa results in the complete sulfidation of Co ions that have diffused into the Al_2O_3 support during the calcination treatment.

Besides MES, the catalysts were characterized using extended X-ray absorption fine structure (EXAFS) to obtain information about the local environments around Co and Mo during sulfidation. The catalytic activity was evaluated by dibenzothiophene (DBT) HDS reaction tests. Complementary information about the influence of the size and morphology of the active structures on the activity was derived from transmission electron microscopy (TEM).

3.2 Experimental

Six $CoMo/Al_2O_3$ catalysts were prepared by pore-volume impregnation of γ-Al_2O_3 (Ketjen 001-1.5E, BET surface area 271 $m^2\ g^{-1}$, pore volume 0.7 ml g^{-1}, particle size 0.5-0.85 mm). Aqueous solutions of cobalt nitrate $Co(NO_3)_2{\cdot}6H_2O$ (Merck p.a.) and ammonium heptamolybdate $(NH_4)_6Mo_7O_{24}{\cdot}4H_2O$ (Merck, min. 99.9%) were used in a two-step impregnation procedure, Mo being introduced first. About 50 MBq ^{57}Co as an aqueous solution of $Co(NO_3)_2 \cdot 6H_2O$ was added to the Co-containing impregnation solution, for the MES measurements.

The resulting $CoMo/Al_2O_3$ catalysts contained 7 wt% Mo, with four high promoter-loading catalysts having 2.25 wt% Co, denoted as $Co(2.25)Mo(7)/Al_2O_3$, and two low promoter-loading catalysts having 0.04 wt% Co, denoted as $Co(0.04)Mo(7)/Al_2O_3$. After preparation, the catalysts were dried in static air at 383 K for 16 h. Subsequently the samples were calcined at 673 K for 24 h in static air. Two high promoter-loading catalysts were afterwards calcined for 2 h at 773 K and 873 K respectively. The calcination temperature is indicated as a postfix.

Prior to Mössbauer spectroscopic measurements the catalysts were sulfided in a flow of 60 cm^3 min^{-1} of a mixture of 10 vol.% H_2S in H_2 at pressures between 0.1 and 4 MPa in a high-pressure Mössbauer *in situ* cell, earlier described in detail [16]. This cell offers the possibility to study catalysts under high-pressure conditions. The sulfidation procedure is denoted by the suffix (S, x MPa, y K), indicating heating of the catalyst in the sulfidation mixture to y K at x MPa in 1 h and keeping the catalyst at this temperature for 1 hour.

The experimental method applied for the MES measurements is presented in section 2.1.4 and the details of the EXAFS measurements and analysis are available in 2.2.2. For the TEM experiments described in section 2.3, the same sulfided samples were used as for the EXAFS measurements. Gas-phase DBT HDS experiments were carried out in a high-pressure stainless steel reactor according to the method described in 2.4.

3.3 Results

3.3.1 MES results

The Mössbauer spectra of the $Co(2.25)Mo(7)/Al_2O_3$-673 catalyst sulfided at 4 MPa are presented in Figure 3.1 and the resulting MES parameters are shown in Table 3.1. The spectrum of the fresh catalyst consists of one quadrupole doublet indicating a low-spin Co 2+ or a 3+ phase and a second contribution consistent with Co^{2+} in the high-spin state [17]. The doublet with an I.S. value of 0.28 mm s^{-1} and a Q.S. of 0.95 mm s^{-1} is assigned to a Co-oxide species. The high-spin 2+ contribution observed in the MES spectra is assigned mainly to Co diffused in the Al_2O_3 support ($Co:Al_2O_3$) during the calcination treatment.

Exposure of the fresh catalyst to the H_2S/H_2 gas mixture has an influence on the catalyst even at room temperature. After treatment at room temperature and atmospheric pressure, the high-spin 2+ phase contribution increases. This behavior was also observed in previous MES studies of uncalcined catalysts with about the same Co and Mo loadings [9]. Upon room temperature sulfidation at 4 MPa, the high-spin doublet slightly decreases, while the Q.S. of the second doublet diminishes considerably. The Q.S. value of this doublet is continuously decreasing after sulfidation up to 473 K and increases after treatment at 573 K.

The high-spin 2+ spectral contribution gradually decreases with increasing sulfidation temperature after 373 K and has disappeared at 673 K in favor of a doublet with I.S. = 0.22 mm s^{-1} and Q.S. = 0.94 mm s^{-1}. This doublet points to the presence of Co-Mo-S species. After sulfidation at 673 K and 4 MPa for 4 h, a small doublet is present with an IS of 0.23 mm s^{-1} and a QS of 0.32 mm s^{-1} that can be attributed to sintered Co-sulfide species.

Figure 3.2 shows the MES spectra obtained with the Co(0.04)Mo(7)/Al_2O_3 catalyst sulfided at 4 MPa and the corresponding fit parameters are presented in Table 3.2. The spectrum of the fresh catalyst shows similar features as that of the Co(2.25)Mo(7)/Al_2O_3-673 catalyst. Upon stepwise sulfidation of Co(0.04)Mo(7)/Al_2O_3 up to 673 K at 4 MPa the high-spin 2+ spectral contribution disappears completely from the spectrum and four hours treatment under the same sulfidation conditions does not change the spectrum further. Again, the Q.S. of the dominant doublet decreases to 0.93 mm s^{-1} during treatments up to 473 K, while it is increasing after sulfidation at 573 K. The resulting doublet has an I.S. value of 0.22 mm s^{-1} and a Q.S. value of 1.22 mm s^{-1}, parameters that are characteristic for Co-Mo-S.

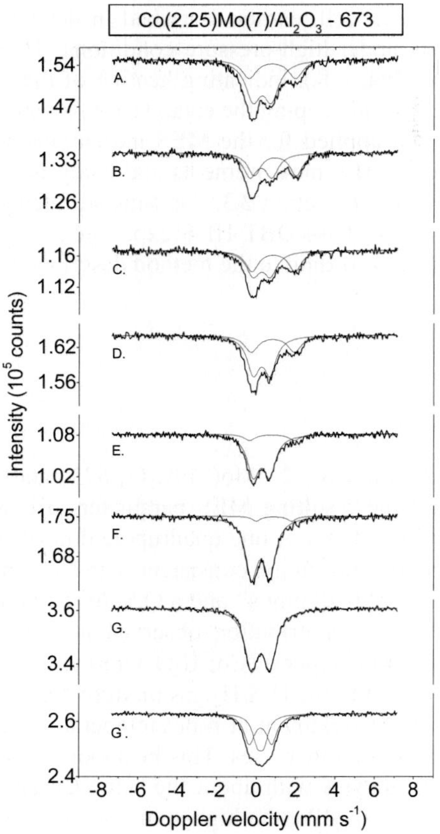

Figure 3.1 *MES spectra obtained at 300 K with Co(2.25)Mo(7)/Al_2O_3-673 catalyst after various successive sulfidation steps. A. Fresh catalyst; B. (S, 0.1 MPa, 300 K); C. (S, 4 MPa, 300 K); D. (S, 4 MPa, 373 K); E. (S, 4 MPa, 473 K); F. (S, 4 MPa, 573 K); G. (S, 4 MPa, 673 K, 1 h); G^*. (S, 4 MPa, 673 K, 4 h).*

Table 3.1 *MES parameters of Co(2.25)Mo(7)/Al_2O_3-673 catalyst after sulfidation treatment*

T_s K	P MPa	IS[a] mm s^{-1}	QS[b] mm s^{-1}	Γ[c] mm s^{-1}	A[d] %	IS mm s^{-1}	QS mm s^{-1}	Γ mm s^{-1}	A %	IS mm s^{-1}	QS mm s^{-1}	Γ mm s^{-1}	A %
		"Co – oxide"				"Co – sulfide"				"High-spin 2+"			
Fresh		0.28	0.95	0.94	68					0.80	2.37	0.88	32
300	0.1					0.22	1.12	0.80	51	0.94	2.15	0.78	49
300	4					0.26	0.89	0.83	54	0.85	2.20	0.79	46
373	4					0.26	0.85	0.76	66	0.83	2.21	0.84	34
473	4					0.26	0.84	0.74	87	0.79	2.28	0.70	13
573	4					0.22	0.93	0.73	94	0.90	1.81	0.51	6
673	4					0.22	0.94	0.78	100				
						"Co – sulfide"				Co_9S_8			
673[e]	4					0.23	1.14	0.66	58	0.23	0.32	0.58	42

[a] Isomer shift: I.S. ± 0.03 mm s^{-1}; [b] Quadrupole splitting: Q.S. ± 0.03 mm s^{-1}; [c] Linewidth: Γ ± 0.05 mm s^{-1};
[d] Spectral contribution: A ± 5%; [e] Treatment at 673 K for four hours.

Table 3.2 *MES parameters of Co(0.04)Mo(7)/Al_2O_3 catalyst after sulfidation treatment*

T_s K	P MPa	IS mm s^{-1}	QS mm s^{-1}	Γ mm s^{-1}	A %	IS mm s^{-1}	QS mm s^{-1}	Γ mm s^{-1}	A %	IS mm s^{-1}	QS mm s^{-1}	Γ mm s^{-1}	A %
		"Co – oxide"				"Co – sulfide"				"High-spin 2+"			
Fresh		0.26	0.70	0.90	45					0.75	2.34	1.38	55
300	0.1					0.23	1.33	1.27	44	0.97	2.15	0.76	56
300	4					0.20	1.32	0.86	37	0.94	2.11	0.80	63
373	4					0.29	0.93	0.83	63	0.87	2.22	0.75	37
473	4					0.28	0.93	0.77	87	0.78	2.41	0.54	13
573	4					0.22	1.13	0.70	95	1.07	2.09	0.40	5
673	4					0.22	1.23	0.64	100				
673[a]	4					0.22	1.22	0.66	100				

[a] Treatment at 673 K for four hours.

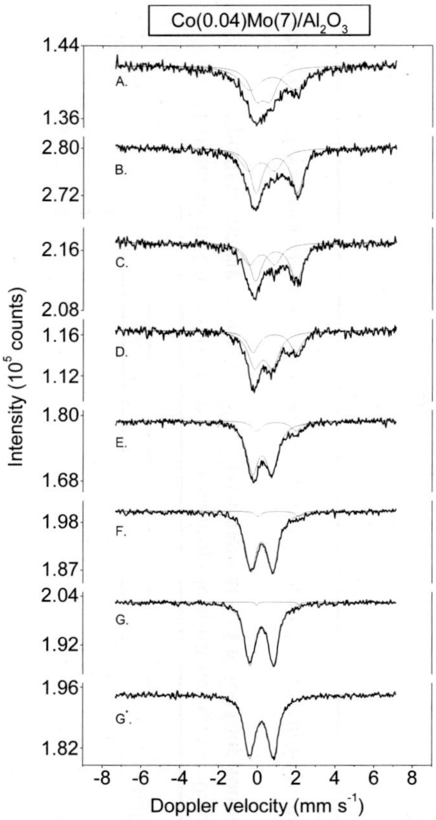

Figure 3.2 *MES spectra obtained at 300 K with Co(0.04)Mo(7)/Al_2O_3 catalyst after various successive sulfidation steps. A. Fresh catalyst; B. (S, 0.1 MPa, 300 K); C. (S, 4 MPa, 300 K); D. (S, 4 MPa, 373 K); E. (S, 4 MPa, 473 K); F. (S, 4 MPa, 573 K); G. (S, 4 MPa, 673 K, 1 h); G^*. (S, 4 MPa, 673 K, 4 h).*

To better understand the influence of the use of elevated pressure during sulfidation of the catalysts, the same samples were sulfided first at atmospheric pressure and then with incremental pressure steps of 1 MPa to 4 MPa. Mössbauer spectra of Co(2.25)Mo(7)/Al_2O_3 in the fresh state, after temperature-programmed atmospheric-pressure sulfidation up to 673 K and after subsequent sulfidation at pressures of 1, 2, 3 and 4 MPa at 673 K are presented in Figures 3.3 and 3.4. The corresponding MES fit parameters are collected in Table 3.3.

To observe the influence of water removal on the high spin 2+ contribution in the fresh catalyst, the sample was dried in an Ar flow at 393 K for 2 h. This treatment resulted in a clear increase of the high spin 2+ contribution. The removal of water results in an increased interaction of such Co species with the support [12]. The high-spin 2+ contribution increases strongly after exposing the catalyst to the sulfidation mixture at 300 K. Increasing the sulfidation temperature results in a continuous decrease of this contribution, although it does not completely disappear after sulfidation at 673 K and 0.1 MPa.

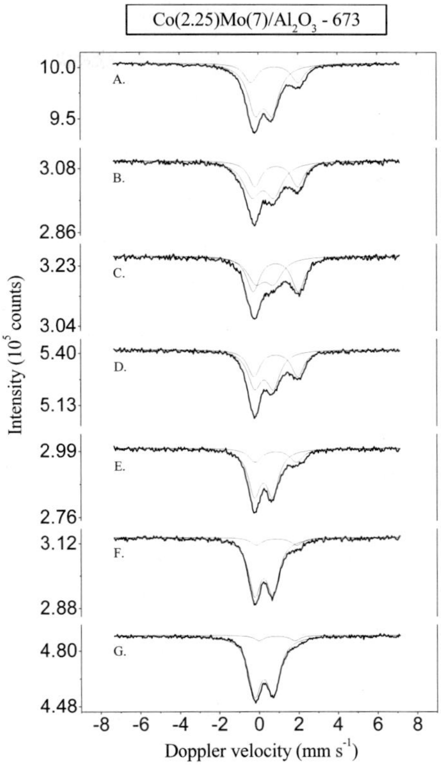

Figure 3.3 *MES spectra obtained at 300 K with Co(2.25)Mo(7)/Al_2O_3-673 catalyst after various successive sulfidation steps. A. Fresh catalyst; B. (Ar, 393 K); C. (S, 0.1 MPa, 300 K); D. (S, 0.1 MPa, 373 K); E. (S, 0.1 MPa, 473 K); F. (S, 0.1 MPa, 573 K); G. (S, 0.1 MPa, 673 K).*

The Q.S. values of the Co-sulfide doublet at intermediate temperatures (Q.S. = 0.95 – 0.99 mm s^{-1}) are somewhat larger than the values previously obtained with the identical catalyst sulfided directly at high-pressure. The Q.S. values decrease during sulfidation from 1.10 mm s^{-1} at 300 K to 0.92 mm s^{-1} at 473 K. A further increase of the sulfidation temperature up to 673 K at 0.1 MPa results in a small increase of the quadrupole splitting. A successive treatment at 673 K and 0.1 MPa showed that prolonged activation under these conditions does not have any influence on the Mössbauer spectra. During the subsequent treatments at elevated pressures, a slight decrease in the Q.S. value of the Co-sulfide doublet is observed. The high-spin 2+ contribution is still present after sulfidation at 673 K and 4 MPa.

The comparative evolution of the Q.S. values of the Co-sulfide doublet after sulfidation of the Co(2.25)Mo(7)/Al_2O_3-673 catalyst at atmospheric and high-pressure is shown in Figure 3.5.

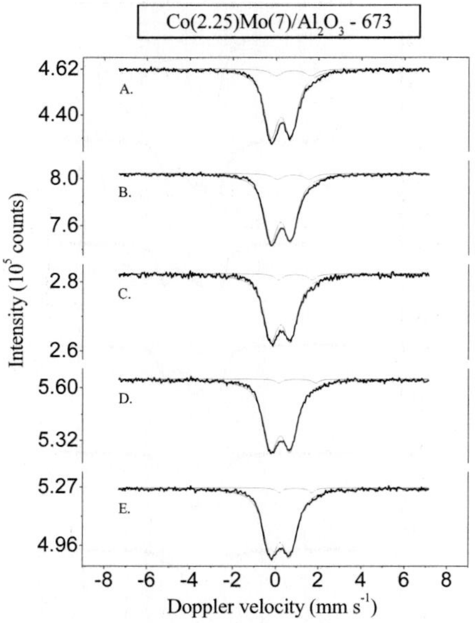

Figure 3.4 *MES spectra obtained at 300 K with Co(2.25)Mo(7)/Al$_2$O$_3$-673 catalyst after various successive sulfidation steps. A. (S, 0.1 MPa, 673 K, repeated); B. (S, 1 MPa, 673 K); C. (S, 2 MPa, 673 K); D. (S, 3 MPa, 673 K); E. (S, 4 MPa, 673 K).*

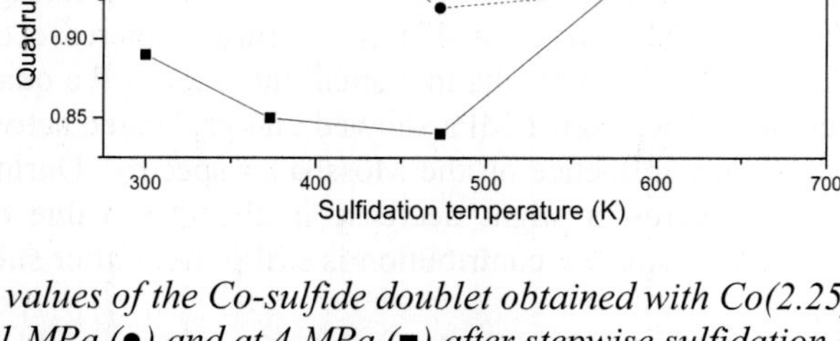

Figure 3.5 *Q.S. values of the Co-sulfide doublet obtained with Co(2.25)Mo(7)/Al$_2$O$_3$-673 catalyst at 0.1 MPa (●) and at 4 MPa (■) after stepwise sulfidation at the indicated temperatures.*

Table 3.3 *MES parameters of Co(2.25)Mo(7)/Al_2O_3-673 catalyst after sulfidation treatment*

T_s K	P MPa	I.S. mm s^{-1}	Q.S. mm s^{-1}	Γ mm s^{-1}	A %	I.S. mm s^{-1}	Q.S. mm s^{-1}	Γ mm s^{-1}	A %	I.S. mm s^{-1}	Q.S. mm s^{-1}	Γ mm s^{-1}	A %
		"Co – oxide"				"Co – sulfide"				"High-spin 2+"			
Fresh		0.27	0.86	0.90	72					0.82	2.39	0.80	28
Ar-393		0.24	1.11	1.06	63					0.90	2.15	0.77	37
300	0.1					0.20	1.10	0.75[b]	38	0.93	2.10	0.89	62
373	0.1					0.24	0.99	0.75[b]	55	0.89	2.12	0.83	45
473	0.1					0.23	0.92	0.75	75	0.86	2.08	0.83	25
573	0.1					0.24	0.93	0.75	88	0.89	2.04	0.72	12
673	0.1					0.24	0.95	0.78	91	0.90	1.81	0.60	9
673[a]	0.1					0.23	0.95	0.77	93	0.82	1.65	0.65	7
673	1					0.23	0.94	0.78	93	0.81	1.50	0.61	7
673	2					0.23	0.92	0.81	93	0.88	1.55	0.60[b]	7
673	3					0.23	0.91	0.83	96	0.96	1.70	0.60[b]	4
673	4					0.23	0.90	0.83	96	0.92	1.54	0.60[b]	4

[a] Repeated treatment at 673 K and 0.1MPa; [b] Fixed during fit.

The MES results obtained with the $Co(0.04)Mo(7)/Al_2O_3$ catalyst sulfided at atmospheric pressure (Figure 3.6) and then subjected to high-pressure treatments (Figure 3.7) are presented in Table 3.4. Sulfidation also takes place via a high-spin 2+ intermediate phase. The spectral contribution of the corresponding doublet clearly increases after the drying treatment in Ar and forms the major contribution to the spectra up to 373 K. The large contribution of high-spin 2+ species hampers a precise determination of the parameters of the second doublet, for which the linewidth values had to be fixed to obtain a reasonable fit. The high-spin 2+ contribution decreases with increasing sulfidation temperature and completely disappears from the spectrum after treatment at 673 K and 2 MPa. The Q.S. values of the Co-sulfide contribution are similar to the values obtained with the same sample sulfided directly at high-pressure, with an evident increase in the Q.S. at 573 K. The evolution of the Q.S. values for the $Co(0.04)Mo(7)/Al_2O_3$ catalyst upon activation under atmospheric and high-pressure conditions is presented in Figure 3.8. No further changes are observed in the MES parameters of this contribution assigned to Co-Mo-S structures, after increasing the sulfidation pressure.

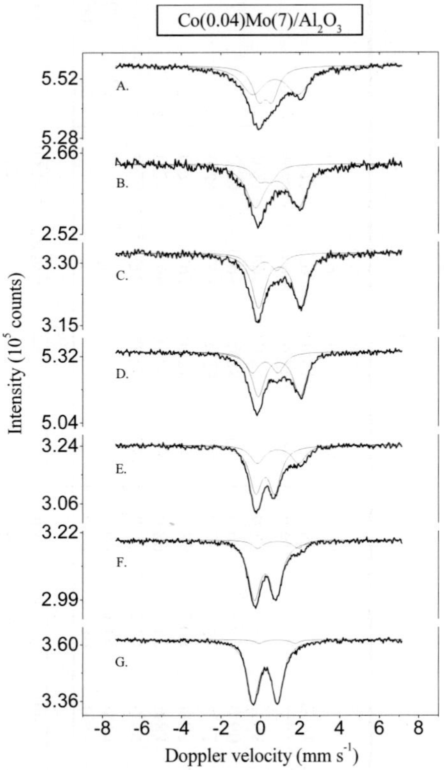

***Figure 3.6** MES spectra obtained at 300 K with $Co(0.04)Mo(7)/Al_2O_3$ catalyst after various successive sulfidation steps. A. Fresh catalyst; B. (Ar, 393 K); C. (S, 0.1 MPa, 300 K); D. (S, 0.1 MPa, 373 K); E. (S, 0.1 MPa, 473 K); F. (S, 0.1 MPa, 573 K); G. (S, 0.1 MPa, 673 K).*

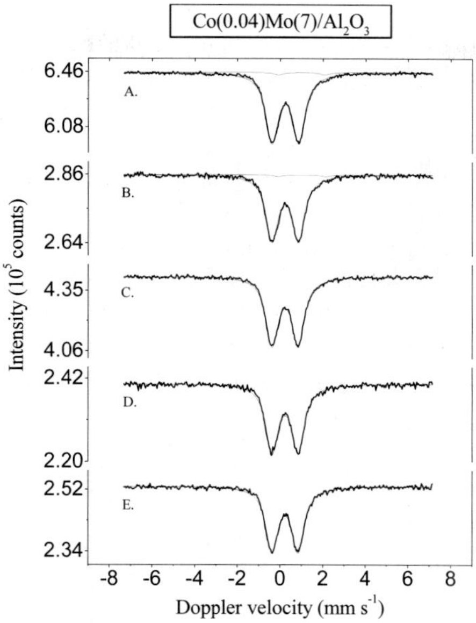

***Figure 3.7** MES spectra obtained at 300 K with Co(0.04)Mo(7)/Al_2O_3 catalyst after various successive sulfidation steps. A. (S, 0.1 MPa, 673 K, repeated); B. (S, 1 MPa, 673 K); C. (S, 2 MPa, 673 K); D. (S, 3 MPa, 673 K); E. (S, 4 MPa, 673 K).*

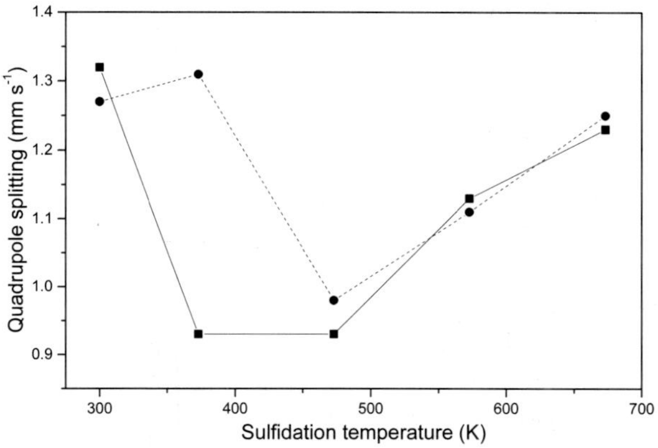

***Figure 3.8** Q.S. values of the Co-sulfide doublet obtained with Co(0.04)Mo(7)/Al_2O_3 catalyst at 0.1 MPa (●) and at 4 MPa (■) after stepwise sulfidation at the indicated temperatures.*

Table 3.4 *MES parameters of $Co(0.04)Mo(7)/Al_2O_3$ catalyst after sulfidation treatment*

T_s K	P MPa	I.S. mm s^{-1}	Q.S. mm s^{-1}	Γ mm s^{-1}	A %	I.S. mm s^{-1}	Q.S. mm s^{-1}	Γ mm s^{-1}	A %	I.S. mm s^{-1}	Q.S. mm s^{-1}	Γ mm s^{-1}	A %
		"Co – oxide"				"Co – sulfide"				"High-spin 2+"			
Fresh		0.26	0.68	0.80[b]	40					0.75	2.30	1.37	60
Ar-393		0.24	0.60	0.80[b]	20					0.83	2.18	1.12	80
300	0.1					0.21	1.27	0.80[b]	23	0.96	2.12	0.83	77
373	0.1					0.24	1.31	0.80[b]	31	0.94	2.13	0.82	69
473	0.1					0.24	0.98	0.76	68	0.88	2.11	0.86	32
573	0.1					0.23	1.11	0.74	88	0.88	2.07	0.66	12
673	0.1					0.24	1.25	0.69	96	0.85	1.84	0.60[b]	4
673[a]	0.1					0.24	1.25	0.70	97	1.00	2.16	0.60[b]	3
673	1					0.24	1.24	0.69	98	1.02	2.20	0.60[b]	2
673	2					0.23	1.22	0.70	100				
673	3					0.24	1.23	0.69	100				
673	4					0.23	1.22	0.69	100				

[a] Repeated treatment at 673 K and 0.1MPa; [b] Fixed during fit.

The results obtained with the two catalysts calcined at higher temperatures, Co(2.25)Mo(7)/Al_2O_3-873 and Co(2.25)Mo(7)/Al_2O_3-773, after stepwise sulfidation at 4 MPa, are presented in Figure 3.9 and Figure 3.10, respectively. The corresponding MES parameters are listed in Table 3.5 and Table 3.6.

The spectra of the fresh catalysts were analyzed in terms of two Co^{3+} and two Co^{2+} doublets, corresponding to octahedrally and tetrahedrally coordinated Co species [18,19]. An increase in the calcination temperature from 773 to 873 K results in the predominant formation of tetrahedrally coordinated Co^{2+} species. The Q.S. value of this spectral contribution decreases with increasing calcination temperature, but the formation of appreciable amounts of the stoichiometric $CoAl_2O_4$ phase does not occur [18].

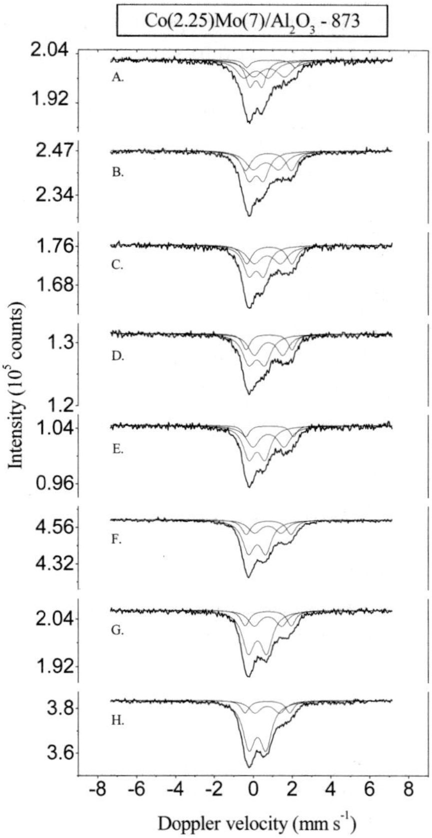

***Figure 3.9** MES spectra obtained at 300 K with Co(2.25)Mo(7)/Al_2O_3-873 catalyst after various successive sulfidation steps. A. Fresh catalyst; B. (S, 0.1 MPa, 300 K); C. (S, 4 MPa, 300 K); D. (S, 4 MPa, 373 K); E. (S, 4 MPa, 473 K); F. (S, 4 MPa, 573 K); G. (S, 4 MPa, 673 K); H. (S, 4 MPa, 773 K).*

To simplify the fitting procedure, the spectra of the sulfided catalysts were analyzed in terms of only one "Co-sulfide" component and two Co^{2+} doublets. Again, the formation of a high-spin 2+ intermediate phase, which is octahedrally coordinated, is observed after exposing the catalysts to the H_2S/H_2 reaction mixture at room temperature. The spectral contribution of the octahedrally and tetrahedrally coordinated Co^{2+} species continuously decreases with increasing sulfidation temperature at 4 MPa, but both doublets are still present even after treatment at 773 K and 4 MPa. A small increase in the Q.S. value of the Co-sulfide doublet is observed for both catalysts after sulfidation at 573 K and 4 MPa. About 70% of the Co is present in a Co-Mo-S phase in the spectrum of the Co(2.25)Mo(7)/Al_2O_3-873 catalyst sulfided at 773 K and 4 MPa. After activation of the Co(2.25)Mo(7)/Al_2O_3-773 catalyst at 673 K and 4 MPa, about 90% of the total Co is sulfided and after increasing the sulfidation temperature to 773 K, a new doublet (I.S. = 0.23 mm s^{-1} and Q.S. = 0.44 mm s^{-1}), assigned to sintered Co-sulfide species, is observed in the spectrum.

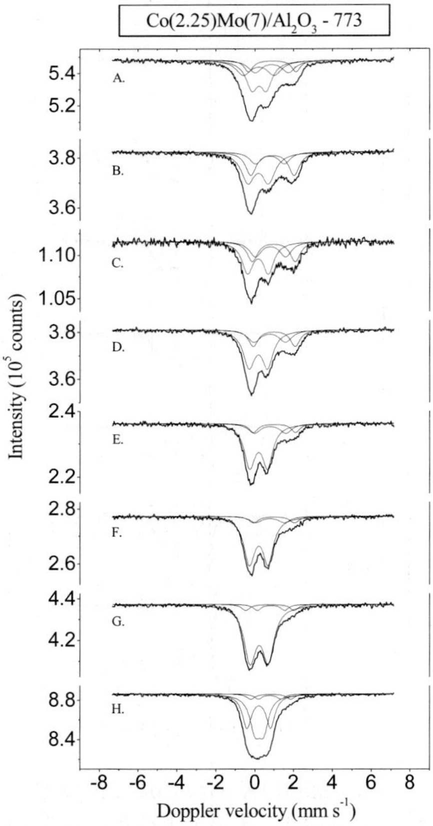

Figure 3.10 *MES spectra obtained at 300 K with Co(2.25)Mo(7)/Al_2O_3-773 catalyst after various successive sulfidation steps. A. Fresh catalyst; B. (S, 0.1 MPa, 300 K); C. (S, 4 MPa, 300 K); D. (S, 4 MPa, 373 K); E. (S, 4 MPa, 473 K); F. (S, 4 MPa, 573 K); G. (S, 4 MPa, 673 K); H. (S, 4 MPa, 773 K).*

***Table 3.5** MES parameters of Co(2.25)Mo(7)/Al_2O_3-873 catalyst after sulfidation treatment*

T_s K	P MPa	IS mm s^{-1}	QS mm s^{-1}	Γ mm s^{-1}	A %	IS mm s^{-1}	QS mm s^{-1}	Γ mm s^{-1}	A %	IS mm s^{-1}	QS mm s^{-1}	Γ mm s^{-1}	A %	IS mm s^{-1}	QS mm s^{-1}	Γ mm s^{-1}	A %
		Co^{3+} component								Co^{2+} component							
		Co_{tet}				Co_{oct}				Co_{tet}				Co_{oct}			
Fresh		0.13	0.62	0.55	28	0.15	1.27	0.92	32	0.80	1.60	1.00	33	0.9	2.48	0.44	7
						"Co – sulfide"				Co_{tet}				Co_{oct}			
300	0.1					0.15	0.73	0.72	38	0.65	1.32	0.98	34	0.78	2.36	0.68	28
300	4					0.16	0.74	0.72	42	0.73	1.36	0.94	34	0.83	2.32	0.62	24
373	4					0.19	0.82	0.79	47	0.79	1.46	0.78	35	0.82	2.36	0.50	18
473	4					0.19	0.82	0.72	50	0.78	1.60	0.85	39	0.85	2.44	0.44	11
573	4					0.20	0.89	0.74	55	0.75	1.37	0.81	25	0.79	2.30	0.57	20
673	4					0.22	0.93	0.75	60	0.76	1.41	0.76	23	0.77	2.37	0.54	17
773	4					0.21	0.87	0.73	68	0.72	1.28	0.74	19	0.73	2.30	0.50	13

***Table 3.6** MES parameters of Co(2.25)Mo(7)/Al_2O_3-773 catalyst after sulfidation treatment*

T_s K	P MPa	IS mm s^{-1}	QS mm s^{-1}	Γ mm s^{-1}	A %	IS mm s^{-1}	QS mm s^{-1}	Γ mm s^{-1}	A %	IS mm s^{-1}	QS mm s^{-1}	Γ mm s^{-1}	A %	IS mm s^{-1}	QS mm s^{-1}	Γ mm s^{-1}	A %
		Co^{3+} component								Co^{2+} component							
		Co_{tet}				Co_{oct}				Co_{tet}				Co_{oct}			
Fresh		0.22	0.75	0.75	40	0.21	1.60	0.92	26	0.88	1.74	0.85	21	0.91	2.46	0.54	13
						"Co – sulfide"				Co_{tet}				Co_{oct}			
300	0.1					0.18	1.07	0.83	50	0.78	1.50	0.73	18	0.91	2.24	0.67	32
300	4					0.17	1.09	0.70	47	0.81	1.57	0.74	25	0.98	2.22	0.64	28
373	4					0.18	0.96	0.72	56	0.77	1.64	0.80	21	1.00	2.17	0.61	23
473	4					0.19	0.92	0.68	70	0.79	1.63	0.73	18	1.02	2.18	0.55	12
573	4					0.20	0.97	0.69	78	0.78	1.48	0.74	12	0.99	2.12	0.57	10
673	4					0.23	0.96	0.78	89	0.82	1.58	0.46	7	0.92	2.63	0.50[a]	4
		Co_9S_8				"Co – sulfide"				Co_{tet}				Co_{oct}			
773	4	0.23	0.44	0.64	46	0.21	1.22	0.57	42	0.77	1.17	0.50	6	0.83	2.07	0.50[a]	6

[a] Fixed during fit.

3.3.2 Mo EXAFS

The fit results of the Mo K-edge EXAFS measurements of Co(2.25)Mo(7)/Al_2O_3-673 as a function of sulfidation temperature at atmospheric and elevated (4 MPa) pressures are collected in Table 3.7.

Table 3.7 *Mo K-edge EXAFS parameters of Co(2.25)Mo(7)/Al_2O_3-673 catalyst after different sulfidation treatments*

Treatment	N -	R Å	$\Delta\sigma^2$ $Å^2$ $\cdot10^{-3}$	ΔE_0 eV	N -	R Å	$\Delta\sigma^2$ $Å^2$ $\cdot10^{-3}$	ΔE_0 eV	N -	R Å	$\Delta\sigma^2$ $Å^2$ $\cdot10^{-3}$	ΔE_0 eV
	Mo-S				Mo-Mo				Mo-Co			
S, 4 MPa, 473 K	6.2	2.47	5.3	1.3	1.9	2.72	6.2	7.9				
S, 4 MPa, 573 K	4.8	2.41	2.5	1.7	2.3	3.15	5.4	4.3	0.8	2.76	9.8	9.8
S, 4 MPa, 673 K	5.6	2.41	1.6	2.1	3.4	3.16	3.7	2.5	0.9	2.75	9.5	7.9
S, 0.1 MPa, 473 K	3.2	2.41	7.3	0	1.0	3.10	9.7	9.3				
S, 0.1 MPa, 573 K	4.0	2.41	3.5	0.7	1.8	3.13	5.8	8.5	0.6	2.78	9.5	5.3
S, 0.1 MPa, 673 K	5.0	2.41	1.8	0.1	3.2	3.16	3.9	0.8	0.7	2.73	9.9	9.5
S, 0.1 MPa + S, 4 MPa, 673 K	5.4	2.41	1.6	1.5	3.4	3.16	3.6	2.2	0.8	2.74	9.9	9.1

The spectrum of the sample sulfided at 473 K and 4 MPa was fitted by two contributions, i.e. a Mo-S contribution at 2.47 Å and a Mo-Mo contribution at 2.72 Å. The Mo-S distance of 2.47 Å is larger than the interatomic Mo-S distance in crystalline MoS_2 (2.41 Å), whereas the Mo-Mo distance is smaller than the typical distance of 3.16 Å in MoS_2 [20]. A relatively high Mo-S coordination number (6.2) is found under these conditions.

After increasing the sulfidation temperature to 573 K and 673 K, the Mo-S and Mo-Mo interatomic distances agree well with the structural parameters of bulk MoS_2, while the observed coordination numbers are lower than those of MoS_2 ($N_{Mo\text{-}S}$ = 6, $N_{Mo\text{-}Mo}$ = 6), indicating the presence of dispersed MoS_2 particles. In contrast, sulfidation at atmospheric pressure results in the presence of species with structural parameters close to those of MoS_2 at all intermediate temperatures. The catalyst activated at 673 K and 4 MPa has a slightly higher Mo-S coordination number than the sample sulfided at 0.1 MPa.

The Mo-Mo coordination numbers and the corresponding Debye-Waller factors are very similar for all catalysts sulfided at 673 K. In all spectra of catalysts sulfided above 473 K, inclusion of a Mo-Co contribution was found to be relevant at distances ranging from 2.74 Å and 2.78 Å.

Table 3.8 *Co K-edge EXAFS parameters of Co(2.25)Mo(7)/ Al_2O_3-673catalyst after different sulfidation treatments*

Treatment	N -	R Å	$\Delta\sigma^2$ $Å^2 \cdot 10^{-3}$	ΔE_0 eV	N -	R Å	$\Delta\sigma^2$ $Å^2 \cdot 10^{-3}$	ΔE_0 eV	N -	R Å	$\Delta\sigma^2$ $Å^2 \cdot 10^{-3}$	ΔE_0 eV	N -	R Å	$\Delta\sigma^2$ $Å^2 \cdot 10^{-3}$	ΔE_0 eV
	Co-S				Co-Co (1)				Co-Mo				Co-Co (2)			
S, 4 MPa, 300 K	5.3	2.22	4.6	2.6	1.4	2.46	9.8	5.9					0.5	3.35	9.8	4.6
S, 4 MPa, 373 K	5.4	2.21	4.4	2.9	1.3	2.49	9.5	0.1					0.8	3.35	9.3	5.2
S, 4 MPa, 473 K	5.6	2.21	4.0	2.5	1.5	2.51	9.7	-3.6					1.0	3.42	9.4	-1.9
S, 4 MPa, 573 K	5.5	2.22	2.9	0.7	0.6	2.51	9.6	-8.9	0.9	2.80	7.4	0.1	0.6	3.48	7.1	-9.9
S, 4 MPa, 673 K	6.1	2.21	3.5	-0.3	0.6	2.51	9.2	-9.6	1.2	2.80	8.6	-0.8	1.1	3.48	10.1	-10
S, 0.1 MPa, 300 K	4.9	2.22	5.7	5.9	1.7	2.43	9.7	8.4					1.3	3.25	11.0	10.0
S, 0.1 MPa, 373 K	5.3	2.22	5.2	2.4	1.8	2.44	9.2	4.7					1.6	3.24	10.7	9.1
S, 0.1 MPa, 473 K	5.4	2.21	4.0	1.0	0.6	2.55	1.3	-10.0					0.9	3.31	10.4	9.6
S, 0.1 MPa, 573 K	5.2	2.22	3.9	1.8	0.8	2.51	9.3	1.6	0.3	2.80	9.2	-7.7	0.7	3.44	10.4	-2.7
S, 0.1 MPa, 673 K	5.6	2.22	2.6	-1.7	0.6	2.6	1.5	-8.9	0.6	2.80	9.2	-9.2	0.4	3.41	4.1	-9.5
S, 0.1 MPa + S, 4 MPa, 673 K	5.7	2.22	2.8	-1.9	0.6	2.6	0.7	-8.5	0.9	2.80	6.2	-8.9	0.3	3.41	3.8	-9.8

3.3.3 Co EXAFS

The Co K-edge EXAFS fit parameters of sulfided Co(2.25)Mo(7)/Al_2O_3-673 are summarized in Table 3.8. The spectra obtained after sulfidation at 300 K and 373 K at elevated pressure could be fitted with a Co-S contribution at a distance of about 2.21 Å and two Co-Co contributions at 2.43-2.49 Å and 3.35 Å. These distances are smaller than the corresponding Co-Co distances in bulk Co_9S_8 (2.51 Å and 3.51 Å, respectively) [21,22]. After increasing the sulfidation temperature to 673 K, both Co-Co distances increase to the values reported for Co_9S_8. A Co-Mo contribution at 2.8 Å was fitted in the spectra measured after treatment at 573 K and 673 K. After sulfidation at 673 K and 4 MPa, the Co-S coordination number increases to 6.1, pointing to a six-fold sulfur coordination by the Co atoms.

The Co EXAFS parameters for CoMo/Al_2O_3 (S, 0.1 MPa, 673 K) show somewhat different Co-Co coordination distances (2.6 Å and 3.41 Å) from those obtained after high-pressure sulfidation. The Co-Mo coordination number after atmospheric pressure sulfidation was 0.6 in contrast to the value of 1.2 found upon sulfidation at 4 MPa. Notably, subsequent sulfidation at 4 MPa of atmospheric pressure sulfided CoMo/Al_2O_3 resulted in an increase of the Co-Mo coordination number ($N_{Co\text{-}Mo}$=0.9).

3.3.4 TEM

Figure 3.11 shows representative TEM micrographs of Co(2.25)Mo(7)/Al_2O_3-673 sulfided at 673 K and various pressures. The micrographs show the well-known MoS_2 slab-like structure. The average slab length and the stacking degree of the three catalysts are summarized in Table 3.9. The slabs appear to be better defined when the catalyst is sulfided at 4 MPa, indicating increased crystallinity after treatment at higher pressure.

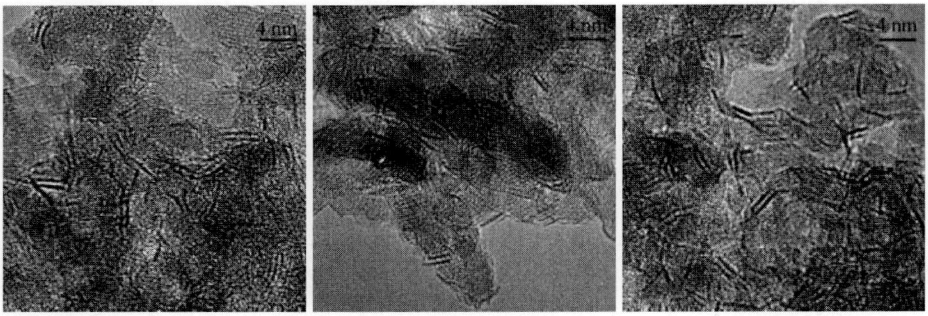

Figure 3.11 *TEM images of Co(2.25)Mo(7)/Al_2O_3-673 catalyst after various sulfidation treatments. Left: (S, 4 MPa, 673 K); middle: (S, 0.1 MPa, 673 K); right: (S, 0.1 MPa, 673 K + S, 4 MPa, 673 K).*

This is evidenced by the more frequent presence of a second direction of *d*-spacings (ordered lines that are visible in a direction perpendicular to the length direction of the slabs). The average slab length as well as the stacking degree does not vary significantly for the three treatments.

Table 3.9 *Slab length and stacking degree of Co(2.25)Mo(7)/ Al_2O_3-673catalyst after different sulfidation treatments*

Treatment	*Average slab length (nm)*	*Average stacking*
S, 4 MPa, 673 K	2.7	1.4
S, 0.1 MPa, 673 K	2.8	1.3
S, 0.1 MPa, 673 K + S, 4 MPa, 673 K	2.9	1.4

3.3.5 Dibenzothiophene HDS

The DBT HDS activities of Co(2.25)Mo(7)/Al_2O_3-673 sulfided at 673 K as a function of sulfidation pressure are shown in Table 3.10. The rate constant (k_{HDS}) is given after a reaction time of 8 h. The apparent activation energies (E_{act}) and the pre-exponential factors (ν_{pre}) were determined in the 533-573 K temperature interval. Clearly, the sulfidation pressure has a very strong influence on the DBT HDS activity of $CoMo/Al_2O_3$: the activity of the catalyst sulfided at 4 MPa is 2.4 times higher than for the sample sulfided under atmospheric conditions. A very similar result was obtained after extending the sulfidation at 4 MPa of an atmospheric pressure sulfided catalyst. Also, a catalyst stepwise sulfided at 4 MPa, thus mimicking the sulfidation procedure of the MES experiments, showed a much higher activity than the catalyst sulfided at 0.1 MPa.

Table 3.10 *DBT HDS activities, apparent activation energies and pre-exponential factors of Co(2.25)Mo(7)/ Al_2O_3-673catalyst after different sulfidation treatments*

Treatment	k_{HDS} *(mol DBT·Kg* $_{Mo}^{-1}\cdot s^{-1}$*)* $\cdot 10^{-2}$	E_{act} *(kJ·mol^{-1})*	ν_{pre} *(mol·kg^{-1}·s^{-1})*
S, 4 MPa, 673 K	4.8	87	$3.7\cdot 10^{6}$
S, 0.1 MPa, 673 K	2.0	79	$3.2\cdot 10^{5}$
S, 0.1 MPa, 673 K + S, 4 MPa, 673 K	4.8	81	$1.2\cdot 10^{6}$
S, 4 MPa, 373 K + S, 4 MPa, 473 K + S, 4 MPa, 573 K + S, 4 MPa, 673 K	4.0	83	$1.5\cdot 10^{6}$

3.4 Discussion

Sulfidation of calcined Co(2.25)Mo(7)/Al_2O_3-673 samples already starts after room-temperature exposure to the sulfidation mixture. The doublet with an I.S. value of 0.22 mm s^{-1} and a Q.S. value of 1.12 mm s^{-1} is assigned to a Co-sulfide phase. Clearly, increasing the sulfidation pressure and temperature causes changes in the shape of the MES spectra.

An increase in the spectral contribution of the high-spin 2+ phase is observed for all samples at this stage and can be explained by removal of water associated with the high-spin species in the fresh catalyst, causing an increased interaction of this species with the support [11]. The formation of an intermediate oxygen containing species during sulfidation, having the same high-spin 2+ doublet – most probably a Co-sulfate type phase – was reported before [8]. The Q.S. of the Co-sulfide doublets decreases to 0.84 mm s^{-1} after high-pressure sulfidation and to 0.92 mm s^{-1} after atmospheric pressure sulfidation at 373 K and 473 K, indicating an increase in the particle size of the Co-sulfide species.

A significant increase in the Q.S. value (from 0.84 mm s^{-1} to 0.93 mm s^{-1}), pointing to a decrease in particle size of the Co-sulfide species, takes place in the sample treated directly at high-pressure (Table 3.1). This can be understood in terms of redispersion of the larger Co-sulfide particles, obtained after the more complete sulfidation at 4 MPa, over the edges of newly formed MoS_2 crystallites. A similar redispersion model has been put forward for NiS_x particles in NiW catalysts [23]. This effect is less pronounced at atmospheric pressure because under such conditions the rate of Co sulfidation is lower and the intermediate CoS_x phase has agglomerated to a lesser extent. Almost no change in the Q.S. value is observed when increasing the temperature from 473 K to 573 K for the 0.1 MPa sulfided catalyst (Table 3.3).

The doublets obtained after sulfidation at 673 K and different pressures with I.S. values of 0.23 and 0.24 mm s^{-1} and Q.S. values of 0.90 and 0.95 mm s^{-1}, are characteristic for Co-Mo-S species [12]. The Mössbauer results indicate that the Co-Mo-S structures are intact upon sulfidation at 673 K under both atmospheric and high-pressure conditions.

The redispersion effect cannot be observed with the Co(0.04)Mo(7)/Al_2O_3 catalyst which tallies with the very well-dispersed nature of the Co-sulfide particles over the edges. Similar Q.S. values are measured upon treatments at 0.1 and 4 MPa (see Figure 3.8). The doublets resulted after sulfidation at 673 K and different pressures have I.S. values of 0.22-0.24 mm s^{-1} and Q.S. values of 1.22-1.25 mm s^{-1}, thus representative of Co-Mo-S species.

During high-pressure sulfidation of CoMo/Al_2O_3 calcined at 673 K the strong interaction between the Co and the Al_2O_3 (characterized by a high-spin 2+ doublet) can be eliminated due to the increased sulfidation rate of Co and all Co atoms end up in the Co-Mo-S phase during the stepwise sulfidation up to 673 K. The Co atoms leave the subsurface positions of the alumina lattice. The observation that they end up at the edges of the MoS_2 slabs in the Co-Mo-S phase suggests considerable mobility. For the first time in a MES study of calcined CoMo/Al_2O_3 catalysts, it is shown that sulfidation at elevated pressure leads to complete sulfidation of Co:Al_2O_3 species. As follows from Figures 3.3 and 3.6, the Co:Al_2O_3 contribution remains in the MES spectra of the samples treated at atmospheric pressure. Even, after a subsequent increase of the sulfidation pressure, Co(2.25)Mo(7)/Al_2O_3-673 shows the presence of a small high-spin 2+ contribution. Seemingly, when these Co ions in strong interaction with the support are not sulfided at low temperature by applying a high sulfidation pressure, they cannot be sulfided at higher temperatures even when the sulfidation pressure is increased at 673 K.
The Co(0.04)Mo(7)/Al_2O_3 catalyst is completely sulfided after the subsequent activation at 673 K and 2 MPa.

After four hours treatment at 673 K and 4 MPa of the Co(2.25)Mo(7)/Al_2O_3-673 catalyst, the Co-sulfide species sintering is observed. The Q.S. value of 0.32 mm s^{-1} of the new-formed doublet is somewhat larger than the Q.S. value of 0.26 mm s^{-1} for crystalline Co_9S_8 [3], and this doublet is assigned to a dispersed Co_9S_8-like species [11]. The longer treatment under these severe sulfidation conditions probably leads to formation of relative large MoS_2 slabs with less edge positions to accommodate all Co atoms and some sintering of Co_9S_8-like particles occurs. The Co_9S_8-like doublet is not observed in the spectrum of the sulfided $Co(0.04)Mo(7)/Al_2O_3$ since the Co_9S_8 segregation is shifted to higher temperatures with decreasing Co content [24].

Previous studies have shown that the interaction of promoter atoms with the alumina increases with increasing the calcination temperature [25], the amount of tetrahedrally coordinated Co growing at the expense of octahedrally coordinated Co [18]. The same trend is observed in this study for $Co(2.25)Mo(7)/Al_2O_3$-873 and $Co(2.25)Mo(7)/Al_2O_3$-773, after stepwise sulfidation at 4 MPa. The amount of Co^{2+} ions in tetrahedral environment increases from about 21 % to 33 % when increasing the calcination temperature from 773 to 873 K. Again, an increase in the spectral contribution of the octahedrally coordinated Co is observed after exposing the catalysts to the H_2S/H_2 gas mixture at room temperature.

To simplify the fitting procedure, the Co^{3+} components were analyzed in terms of only one "Co-sulfide" contribution. The isomer shift values of this contribution are relatively low (Tables 3.5 and 3.6) after treatments at the intermediate temperatures, indicating the presence of tetrahedrally coordinated ions. The doublets resulted after sulfidation at 673 K and 4 MPa have I.S. values of about 0.22 mm s^{-1} and Q.S. values of 0.93-0.96 mm s^{-1}, being assigned to Co-Mo-S species. An increase in the particle size and/or ordering of the Co-sulfide species takes place after treating the $Co(2.25)Mo(7)/Al_2O_3$-873 catalyst at 773 K and 4 MPa, whereas sintering of the Co_9S_8-like species is observed for the $Co(2.25)Mo(7)/Al_2O_3$-773 catalyst sulfided in the same conditions.

The spectral contributions of the octahedrally and tetrahedrally coordinated Co^{2+} species continuously decrease with increasing sulfidation temperature at 4 MPa, but both doublets are still present after sulfidation at 773 K and 4 MPa. At the same time, the Q.S. values of the tetrahedrally coordinated Co^{2+} ions decrease, but the value that is characteristic for $CoAl_2O_4$ is not reached [18].

About 90% of the total Co in $Co(2.25)Mo(7)/Al_2O_3$-773 is sulfided after activation at 673 K and 4 MPa, whereas only 70% of the Co is present in a Co-Mo-S phase after sulfidation at 773 K and 4 MPa. The amount of Co that can be sulfided under high-pressure conditions is higher than previously observed with calcined $CoMo/Al_2O_3$ catalysts sulfided at atmospheric pressure [10,18].

The Co K-edge EXAFS fit results for $Co(2.25)Mo(7)/Al_2O_3$-673 (Table 3.8) show that the Co atoms are sulfided already after exposure to the H_2S/H_2 gas mixture at room temperature. The resulting Co-S coordination distances are similar to the values reported for Co_9S_8 [8, 22, 26]. The two Co-Co contributions at 2.51 Å and at 3.48 Å of the catalysts sulfided at high pressure are closer to the corresponding distances in Co_9S_8 compared to the catalyst sulfided first at atmospheric pressure, in which coordination distances at 2.6 and 3.41 Å are found. This confirms the formation of larger Co-sulfide species upon high-pressure activation as a result of the higher rate of Co sulfidation under these conditions.

The Co-S coordination numbers increase both during atmospheric and high-pressure sulfidation. A coordination number of about 6 is obtained at a sulfidation temperature of 673 K. This forms an indication that no large amounts of Co_9S_8 are present at this stage. Also the relatively small Co-Co coordination numbers support such conclusion. Similar results were obtained for CoMo/C [8] and $NiMo/SiO_2$ catalysts [29].

The formation of an intermediate sulfided species, with very disordered Co-S structures and coordination distances and numbers different from those of crystalline cobalt sulfide systems, was proposed. These special structures are formed as soon as the catalysts are sulfided and only after activation at 573 and 673 K the Co atoms will participate in the Co-Mo-S species or form Co_9S_8. The stronger redispersion of the Co-sulfide particles after direct high-pressure sulfidation at 573 K is confirmed by the significant decrease in the Co-Co contribution from $N_{Co\text{-}Co} = 1.5$ (after treatment at 473 K) to $N_{Co\text{-}Co} = 0.6$ (Table 3.8). No decrease of the Co-Co coordination number at 573 K was observed for the sample treated under atmospheric pressure conditions.

The EXAFS data analyses of the catalysts sulfided at 573 K reveal the presence of a Co-Mo contribution at 2.8 Å, in good agreement with the previously reported values [8,28]. This is in line with the conclusion that Co-Mo-S structures are formed at this temperature. Higher Co-Mo coordination numbers were found for the catalysts activated at 4 MPa. This trend points in direction of Type II Co-Mo-S phase formation upon increasing sulfidation pressure. The Co species become better defined upon sulfidation at higher pressure.

The Mo EXAFS fit results for Co(2.25)Mo(7)/Al_2O_3-673 sulfided directly at high pressure (Table 3.7) reveal the presence of a large Mo-S coordination number ($N_{Mo\text{-}S} = 6.2$), in combination with a short Mo-Mo distance (2.72 Å), after treatment at 473 K. Previous EXAFS studies attributed these contributions to a MoS_3-like intermediate species [29,30,31]. After increasing the sulfidation temperature to 573 K and 673 K, the Mo-S and Mo-Mo interatomic distances are in agreement with the structural parameters of MoS_2 crystallites. The sample sulfided at atmospheric pressure shows the presence of MoS_2-like species at all intermediate temperatures. A lower Mo-S coordination number ($N_{Mo\text{-}S} = 5.0$) is obtained after treatment at 673 K and 0.1 MPa, indicating that the MoS_2 edges are not completely saturated with sulfur, probably due to the presence of some Mo-O-Al linkages of Mo atoms with the alumina support or at least incomplete sulfur coordination [32]. This agrees with the definition of the Type I Co-Mo-S phase, which typically forms under these sulfidation conditions. Contributions from Mo-Co coordination are also observed starting from sulfidation at 573 K, although the Mo-Co coordination numbers are slightly smaller compared to those of the catalyst sulfided at 4 MPa. This trend was also observed in the Co EXAFS fit results.

After sulfidation at 673 K and 4 MPa, the Mo-S and Mo-Mo coordination numbers have increased compared to atmospheric pressure sulfidation, indicating an accelerated Mo sulfidation under high-pressure conditions. The resulting Mo-Mo Debye-Waller factors are slightly smaller than those of the catalyst sulfided under atmospheric pressure conditions, pointing to a higher static order. All this data suggest the formation of the Type II Co-Mo-S phase upon high-pressure activation of the catalysts.

The TEM micrographs of Co(2.25)Mo(7)/Al_2O_3-673 sulfided at 673 K and different pressures (Figure 3.11) do not show significant differences in the dispersion or the morphology of MoS_2 slabs. This result is different from the one of Kooyman et al. [13], who found an increase in the slab length after 1 MPa sulfidation compared to sulfidation at ambient pressure for a commercial CoMo/Al_2O_3 catalyst. Slabs appear to be better defined when the catalyst is sulfided also at elevated pressure. This would indicate a better crystallinity of the MoS_2 phase upon a higher sulfidation pressure.

The DBT HDS data of Co(2.25)Mo(7)/Al_2O_3-673 sulfided at 673 K and different pressures (Table 3.10) show reaction rate constants that are 2.4 times higher for the samples treated also at 4 MPa compared to the sample that was activated only at atmospheric pressure. This ratio is somewhat higher than the relative activities of the Type I and II Co-Mo-S species observed in gas phase thiophene HDS, where Type II phase was twice as active than Co-Mo-S I [33]. The somewhat higher degree of Co sulfidation upon sulfidation at 4 MPa may explain this difference.

The apparent activation energies are slightly higher for the catalysts activated under high-pressure conditions, behavior understood in terms of increasing surface coverage by dibenzothiophene or H_2S [34]. A strong interaction of the active species with the adsorbed complex can also explain the high pre-exponential factors obtained with these samples. The large changes in the pre-exponential factors could also be related to differences in the number of active sites formed [35].

The large difference in activity between the catalysts sulfided at atmospheric and elevated (4 MPa) pressures is remarkable, the more since our spectroscopic results including those probing the environments around Co (MES, EXAFS) and Mo (EXAFS) do not probe any significant differences in the final state. However, the stepwise approach taken in this study helps to understand evolution of the Co-Mo-S phase in detail. The higher Co-Mo coordination numbers and the more crystalline MoS_2 slabs for the sample obtained after 4 MPa sulfidation give support for the formation of Type II Co-Mo-S structures.

The finding that sulfidation at elevated pressure leads to predominantly Type II Co-Mo-S formation is an important one. The current state of knowledge is that one requires modified preparation routes, for instance using chelating agents [29,36], to diminish metal-support interactions during catalyst activation so as to arrive at fully sulfided Type II structures with a reasonably high dispersion. Considerable improvements in the performance of commercial CoMo- and NiMo-based hydrotreating catalysts have been attributed to maximization of the amount of Type II phase. Despite widely varying activation procedures in industry, in general catalysts are brought into the sulfided state under elevated pressure conditions. As evidenced in this study, a typical calcined CoMo/Al_2O_3 catalyst, which as expected performs as a Type I catalyst following atmospheric pressure sulfidation, renders a much more active, Type II catalyst upon high-pressure activation, albeit at considerably higher temperatures than generally applied commercially. Nevertheless, this observation warrants a re-evaluation of the previous statements and a closer look at the question to what extent it is possible to preserve Type I Co-Mo-S under industrial sulfidation conditions.

3.5 Conclusions

The sulfidation of calcined CoMo/Al_2O_3 catalysts under conditions relevant to industrial practice was studied by MES, EXAFS, TEM and DBT HDS. The amount of Co atoms that participate in Co-Mo-S structures increases upon activation at elevated (4 MPa) pressure. The stepwise sulfidation approach points to higher sulfidation rates of Co and Mo during high-pressure activation. This leads to somewhat more aggregated CoS_x species at intermediate temperatures, followed by redispersion of these species to more dispersed ones, once MoS_2 is formed. The final state of the catalyst after 0.1 MPa or 4 MPa sulfidation is very similar in terms of Q.S. of the Co-Mo-S phase. Nonetheless, the high-pressure sulfided sample has a much higher HDS activity than the sample activated at 0.1 MPa. The difference is due to formation of Type II Co-Mo-S phase at elevated pressures.

Spectroscopically, support for this is found in increased Co-Mo coordination numbers and increased crystallinity of the MoS_2 slabs in transmission electron microscopy.

When the catalysts calcined at 673 K are directly sulfided at 4 MPa, all Co species including those strongly interacting with the alumina support can be sulfided. On the contrary, increasing the sulfidation pressure to 4 MPa subsequent to sulfidation at 673 K and 0.1 MPa leaves part of the Co atoms in strong interaction with alumina. This latter procedure also results in a Type I to Type II Co-Mo-S phase transformation and higher HDS activity. These results suggest that it is difficult to preserve Type I Co-Mo-S structures under industrial sulfidation conditions.

References

1. Directive 2003/17/EC of the European Parliament and of the Council on the Quality of Petrol and Diesel Fuels, Brussels, 3 March 2003.
2. B.S. Clausen, S. Mørup, H. Topsøe and R. Candia, J. Phys. 37 (1976) 249.
3. H. Topsøe, B.S. Clausen, R. Candia, C. Wivel and S. Mørup, J. Catal. 68 (1981) 433.
4. C. Wivel, R. Candia, B.S. Clausen, S. Mørup and H. Topsøe, J. Catal. 68 (1981) 453.
5. A.I. Dugulan, M.W.J. Crajé and G.J. Kearley, J. Catal. 222 (2004) 281.
6. A.I. Dugulan, M.W.J. Crajé, A.R. Overweg and G.J. Kearley, AIP Conf. Proc. 765 (2005) 26.
7. M.W.J. Crajé, V.H.J. de Beer and A.M. van der Kraan, Appl. Catal. 70 (1991) L7-L13.
8. M.W.J. Crajé, S.P.A. Louwers, V.H.J. de Beer, R. Prins and A.M. van der Kraan, J. Phys. Chem. 96 (1992) 5445.
9. M.W.J. Crajé, V.H.J. de Beer, J.A.R. van Veen and A.M. van der Kraan, J. Catal. 143 (1993) 601.
10. J.A.R. van Veen, E. Gerkema, A.M. van der Kraan and A. Knoester, J. Chem. Soc. Chem. Com. (1987) 1684.
11. M.W.J. Crajé, V.H.J. de Beer, J.A.R. van Veen and A.M. van der Kraan, Appl. Catal. A 100 (1993) 97.
12. M.W.J. Crajé, Ph.D. Thesis, Delft University of Technology, Delft, 1992, ISBN 90-73861-08-X.
13. P.J. Kooyman, J.G. Buglass, H.R. Reinhoudt, A.D. van Langeveld, E.J.M. Hensen, H.W. Zandbergen and J.A.R. van Veen, J. Phys. Chem. B 106 (2002) 11795.
14. H. Topsøe, Stud. Surf. Sci. Catal. 130 (2000) 1.
15. H. Topsøe, J. Catal. 216 (2003) 155.
16. M.W.J. Crajé, A. M. Van Der Kraan, J. van de Loosdrecht and P. J. van Berge, Catal. Today 71 (2002) 369.
17. J.G. Steven, L. Zhe, H. Pollak, V.E. Stevens, R.H. White and J.L. Gibson, Mössbauer Handbook Minerals, Mössbauer Effect Data Center, University of North Carolina, Asheville, 1983.
18. C. Wivel, B.S. Clausen, R. Candia, S. Mørup and H. Topsøe, J. Catal. 87 (1984) 497.

19. J.A.R. van Veen, E. Gerkema, A.M. van der Kraan, P.A.J.M. Hendriks and H. Beens, J. Catal. 133 (1992) 112.
20. T.G. Parham, R.P. Merrill, J. Catal. 85 (1984) 295.
21. S.M.A.M. Bouwens, F.B.M. van Zon, M.P. van Dijk, A.M. van der Kraan, V.H.J. de Beer, J.A.R. van Veen, D.C. Koningsberger, J. Catal. 146 (1994) 375.
22. S.M.A.M. Bouwens, J.A.R. van Veen, D.C. Koningsberger, V.H.J. de Beer, R. Prins, J. Phys. Chem. 95 (1991) 123.
23. E.J.M. Hensen, Y. van der Meer, J.A.R. van Veen and J.W. Niemantsverdriet, Appl. Catal. A, 322 (2007) 16.
24. R. Candia, B.S. Clausen, H. Topsøe, Effect of Sulfiding Temperature on Activity and Structure of Co-Mo/Al_2O_3 Catalysts, Proc. 9th Iberoamerican Symp. Catal., 16-21 July 1984, Lisbon, Portugal, 1984, p. 211.
25. Y. Okamoto, S. Ishihara, M. Kawano, M. Satoh and T. Kubota, J. Catal. 217 (2003) 12.
26. W. Niemann, B.S. Clausen, H. Topsøe, Catal. Lett. 4 (1990) 355.
27. L. Medici, R. Prins, J. Catal. 163 (1996) 38.
28. S.M.A.M. Bouwens, D.C. Koningsberger, V.H.J. de Beer, S.P.A. Louwers, R. Prins, Catal. Lett. 5 (1990) 273.
29. R. Cattaneo, T. Weber, T. Shido, R. Prins, J. Catal. 191 (2000) 225.
30. S.P. Cramer, K.S. Liang, A.J. Jacobson, C.H. Chang, R.R. Chianelli, Inorg. Chem. 23 (1984) 215.
31. R. Cattaneo, F. Rota, R. Prins, J. Catal. 199 (2001) 318.
32. D. Nicosia, R. Prins, J. Catal. 231 (2005) 259.
33. J.A.R. van Veen, H.A. Colijn, P.A.J.M. Hendriks, A.J. van Welsenes, Fuel. Proc. Technol. 35 (1993) 137.
34. E. J. M. Hensen, M. J. Vissenberg, V. H. J. de Beer, J. A. R. van Veen, R. A. van Santen, J. Catal. 163 (1996) 429.
35. E. J. M. Hensen, Ph.D. Thesis, Eindhoven University of Technology, Eindhoven, 2000, ISBN 90-386-2871-4.
36. E.J.M. Hensen, V.H.J. de Beer, J.A.R. van Veen, R.A. van Santen, Catal. Lett. 84 (2002) 59.

Chapter 4

Influence of high-pressure on the sulfidation behavior of uncalcined $CoMo/Al_2O_3$ catalysts

Abstract

Structure-activity correlations are determined for uncalcined $CoMo/Al_2O_3$ catalysts by combining Mössbauer emission spectroscopy (MES), extended X-ray absorption fine structure (EXAFS), transmission electron microscopy (TEM) and dibenzothiophene hydrodesulfurization (HDS) activity measurements. The absence of the calcination step after loading of the calcined Mo/Al_2O_3 catalyst with Co leads to a significant formation of highly dispersed Co_9S_8-type species in the high promoter-loading catalyst. Although the different Co-sulfides present can have similar Mössbauer parameters, MES can still indicate the presence of CoS_x species involving variable number of Co atoms. The stepwise sulfidation approach points to a higher rate of Co and Mo sulfidation at elevated pressure. The EXAFS and the TEM data suggest the formation of the Type II Co-Mo-S phase upon activation at 4 MPa, whereas 0.1 MPa sulfidation results in the less active Type I phase.

4.1 Introduction

The environmental demand for the full introduction of "zero sulfur" gasoline and diesel fuels – having no more than 10 mg/kg (ppm) sulfur content – requires the development of highly active hydrodesulfurization (HDS) catalysts. Topsøe and co-workers introduced a fairly clear picture of the active phase in sulfided CoMo catalysts: the Co-Mo-S phase, in which Co atoms are located at the edges of MoS_2 slabs [1-3]. Two types of Co-Mo-S species were introduced, a Type I, in which some Mo-O-Al linkages remain unsulfided, and a Type II, in which all those linkages have disappeared (so that only a van der Waals interaction between the mixed sulfide and the support remains) [4].

The catalytic performance of the CoMo-based catalysts strongly depends on the preparation conditions [5-7], the calcination temperature being one of the important variables [8,9]. Previous studies have shown that a high calcination temperature will cause Co atoms to migrate into the support, leading to the formation of nonsulfidable Co species [10]. The calcination treatment also modifies the Mo-support interaction and the dispersion and morphology of the Mo sulfide phase upon sulfidation of the oxidic precursor [11]. The number of active sites depends on the dispersion of the MoS_2 slabs, with smaller slabs having a higher number of active edge and corner sites [12].

Although different characterization techniques such as extended X-ray absorption fine structure (EXAFS) [13-15], Mössbauer spectroscopy [16,17], X-ray photoelectron spectroscopy (XPS) [18] and scanning tunneling microscopy (STM) [19] revealed the structure of the active sites of sulfided CoMo catalysts in almost atomic detail, many details about the nature and stability of the active sites have not yet been elucidated, especially for catalysts activated under industrial conditions. In this thesis the industrial conditions are approximated by sulfiding the catalysts in a H_2S/H_2 mixture at higher pressures than atmospheric (i.e. 4 MPa).

To better understand the influence of calcination conditions on the performance of $CoMo/Al_2O_3$ catalysts, we prepared a set of catalysts by sequential impregnation of Mo (followed by calcination) and Co (followed by drying). The formation and evolution of the active phase after the different applied treatments were followed using Mössbauer Emission Spectroscopy (MES) and EXAFS. The characterization of the catalysts was completed with high-resolution TEM and catalytic activity measurements (DBT HDS).

4.2 Experimental

Four ^{57}Co-containing CoMo catalysts were prepared by pore volume impregnation of γ-Al_2O_3, according to the method described in section 3.2.1. The four $CoMo/Al_2O_3$ catalysts contained 7 wt% Mo. Two catalysts were prepared with high promoter loading (2.25 wt% Co, denoted as Co(2.25)Mo(7)/Al_2O_3) and two more catalysts with lower Co loadings (1.3 wt% Co, denoted as Co(1.3)Mo(7)/Al_2O_3). The catalyst loadings are given relative to the support material and are calculated from the impregnation solutions.

After Mo introduction, the catalysts were dried in static air at 383 K for 16 h and afterwards heated to 673 K in 5 h and calcined at this temperature for 24 h in static air. After Co introduction the catalysts were only dried in static air at 383 K for 16 h. The drying temperature is also indicated in the catalyst notation to indicate the absence of a subsequent calcination treatment. Before activation, the catalysts were subjected to a drying treatment in hydrogen gas at 393 K for 24 hours in order to remove residual nitrate groups and to observe the influence of the removal of water associated with the high-spin Co 2+ species on the interaction of these structures with the support [20].

The samples used for MES measurements were sulfided in a flow of 60 cm^3 min^{-1} of 10% H_2S/H_2 mixture at pressures between 0.1 and 4 MPa in a high-pressure Mössbauer *in-situ* reactor, similar to the reactor described in detail in [21]. This reactor offers the possibility to study the catalysts under realistic conditions. The applied sulfidation treatment is denoted (S, x MPa, y K), indicating that during the experiment the catalyst is linearly heated to y K at x MPa in 1 h and kept at that temperature for 1 h.

The experimental method applied for the MES measurements is presented in section 2.1.4 and the details of the EXAFS measurements and analysis are available in section 2.2.2. For the TEM experiments described in section 2.3, the same sulfided samples were used as for the EXAFS measurements. Gas-phase DBT HDS experiments were carried out in a high-pressure stainless steel reactor according to the method described in 2.4.

4.3 Results

4.3.1 MES results

The Mössbauer spectra of the Co(2.25)Mo(7)/Al_2O_3-383 catalyst sulfided at 4 MPa are presented in Figure 4.1 and the resulting MES parameters are shown in Table 4.1. The spectrum of the fresh catalyst consists of one quadrupole doublet assigned to a low-spin 2+ or a 3+ phase and a second contribution consistent with Co^{2+} in the high-spin state [22]. The low resonant absorption area (RAA) in the poorly resolved spectrum of the fresh catalyst is explained in terms of large quantities of adsorbed water, which decreases the effective recoil-free fraction. The high-spin 2+ contribution observed in the MES spectra is assigned mainly to Co interacting with the Al_2O_3 support.

After the drying treatment, but also after exposing the catalyst to the H_2S/H_2 gas mixture at room temperature and atmospheric pressure, the high-spin 2+ phase contribution increases. Similar behavior was observed earlier with calcined catalysts (Chapter 3). Upon room temperature sulfidation at 4 MPa, the high-spin doublet increases further, while the Q.S. of the second doublet – assigned to a Co-sulfide species – is decreasing. The Q.S. value of this doublet is continuously decreasing upon further sulfidation up to 473 K and slightly increases after treatment at 573 K.

The high-spin 2+ spectral contribution gradually decreases with increasing sulfidation temperature after 373 K and has disappeared at 473 K in favor of a Co-sulfide doublet with I.S. = 0.25 mm s^{-1} and Q.S. = 0.82 mm s^{-1}. After treatment at 723 K and 4 MPa, a small doublet is present with an IS of 0.25 mm s^{-1} and a QS of 0.44 mm s^{-1} that can be attributed to sintered Co-sulfide species, while the other component now has a Q.S. value of 1.18 mm s^{-1}.

To better understand the influence of high-pressure on the sulfidation behavior of the catalysts, the same catalyst was sulfided first at atmospheric pressure to 673 K and then with increasing pressures up to 4 MPa at this temperature. The corresponding MES spectra are collected in Figures 4.2 and 4.3. The fit parameters are listed in Table 4.2.

***Table 4.1** MES parameters of Co(2.25)Mo(7)/Al$_2$O$_3$-383 catalyst after sulfidation treatment*

T_s K	P MPa	IS mm s^{-1}	QS mm s^{-1}	Γ mm s^{-1}	A %	IS mm s^{-1}	QS mm s^{-1}	Γ mm s^{-1}	A %	IS mm s^{-1}	QS mm s^{-1}	Γ mm s^{-1}	A %
		"Co – oxide"				"Co – sulfide"				"High-spin 2+"			
Fresh		0.11	0.99	0.97	74					1.15	1.90	0.76	26
H_2/393		0.24	1.26	1.05	57					0.97	2.23	0.71	43
300	0.1					0.23	1.34	0.89	43	0.98	2.17	0.72	57
300	4					0.22	1.22	0.88	51	0.96	2.20	0.66	49
373	4					0.24	0.85	0.68	87	1.00	2.10	0.48	13
473	4					0.25	0.82	0.66	100				
573	4					0.25	0.84	0.69	100				
673	4					0.24	0.84	0.76	100				
						"Co – sulfide"				Co_9S_8			
723	4					0.24	1.18	0.56	44	0.25	0.44	0.63	56

***Table 4.2** MES parameters of Co(2.25)Mo(7)/Al$_2$O$_3$-383 catalyst after sulfidation treatment*

T_s K	P MPa	IS mm s^{-1}	QS mm s^{-1}	Γ mm s^{-1}	A %	IS mm s^{-1}	QS mm s^{-1}	Γ mm s^{-1}	A %	IS mm s^{-1}	QS mm s^{-1}	Γ mm s^{-1}	A %
		"Co – oxide"				"Co – sulfide"				"High-spin 2+"			
H_2/393		0.21	1.33	1.05	67					0.96	2.20	0.68	33
300	0.1					0.22	1.35	1.00	46	0.97	2.20	0.64	54
373	0.1					0.22	1.24	0.88	53	0.96	2.20	0.61	47
473	0.1					0.23	0.88	0.69	97	0.95	1.94	0.60	3
573	0.1					0.24	0.88	0.69	100				
673	0.1					0.25	0.89	0.74	100				
673	0.1					0.25	0.89	0.75	100				
673	1					0.25	0.88	0.77	100				
673	2					0.24	0.88	0.76	100				
673	3					0.24	0.87	0.78	100				
673	4					0.24	0.86	0.78	100				
						"Co – sulfide"				Co_9S_8			
723	4					0.24	1.15	0.62	50	0.25	0.48	0.63	50
773	4					0.25	1.14	0.62	42	0.25	0.35	0.63	58

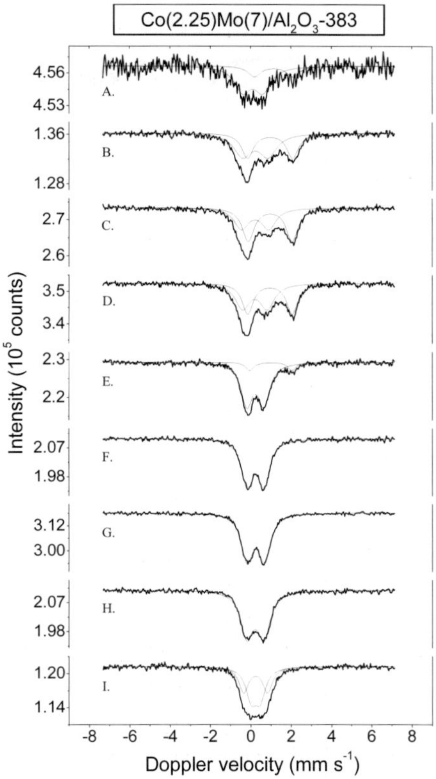

Figure 4.1 *MES spectra obtained at 300 K with Co(2.25)Mo(7)/Al_2O_3-383 catalyst after various successive sulfidation steps. A. Fresh catalyst; B. Drying treatment; C. (S, 0.1 MPa, 300 K); D. (S, 4 MPa, 300 K); E. (S, 4 MPa, 373 K); F. (S, 4 MPa, 473 K); G. (S, 4 MPa, 573 K); H. (S, 4 MPa, 673 K); I. (S, 4 MPa, 723 K).*

Room temperature exposure of the catalyst to the sulfidation mixture leads to a similar MES spectrum as obtained after high-pressure sulfidation. This underpins the reproducibility of the preparation method of the ^{57}Co containing catalysts. The high-spin 2+ contribution does not completely disappear from the spectrum after treatment at 473 K and 0.1 MPa.

The Q.S. values of the second spectral contribution are somewhat larger than the values obtained for the catalyst sulfided at high-pressure. Moreover, the QS value is more constant during the temperature programmed sulfidation procedure at atmospheric pressure. The evolution of the Q.S. values of the Co-sulfide doublet after sulfidation of the Co(2.25)Mo(7)/Al_2O_3-383 catalyst at atmospheric and high-pressure is compared in Figure 4.4. Before increasing the sulfidation pressure, the treatment at 673 K and 0.1 MPa was repeated in order to see whether a longer activation under similar conditions would have any influence on the spectrum. The corresponding spectra in Figures 4.2 and 4.3 and the fit parameters indicate that this was not the case. Similar to the above-noted trends, the high-pressure treatment at 723 K leads to a significant contribution of a doublet with an IS of 0.25 mm s^{-1}, assigned to sintered Co-sulfide species.

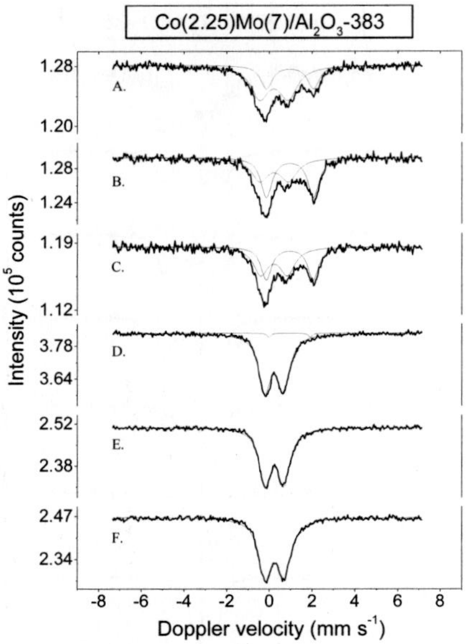

Figure 4.2 *MES spectra obtained at 300 K with Co(2.25)Mo(7)/Al$_2$O$_3$-383 catalyst after various successive sulfidation steps. A. Drying treatment; B. (S, 0.1 MPa, 300 K); C. (S, 0.1 MPa, 373 K); D. (S, 0.1 MPa, 473 K); E. (S, 0.1 MPa, 573 K); F. (S, 0.1 MPa, 673 K).*

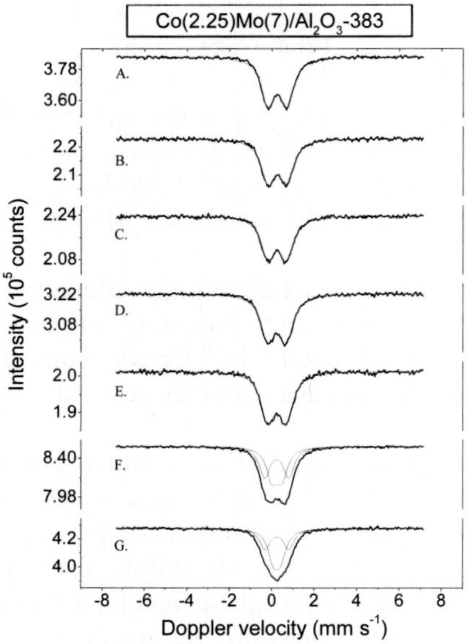

Figure 4.3 *MES spectra obtained at 300 K with Co(2.25)Mo(7)/Al$_2$O$_3$-383 catalyst after various successive sulfidation steps. A. (S, 0.1 MPa, 673 K, repeated); B. (S, 1 MPa, 673 K); C. (S, 2 MPa, 673 K); D. (S, 3 MPa, 673 K); E. (S, 4 MPa, 673 K); F. (S, 4 MPa, 723 K); G. (S, 4 MPa, 773 K).*

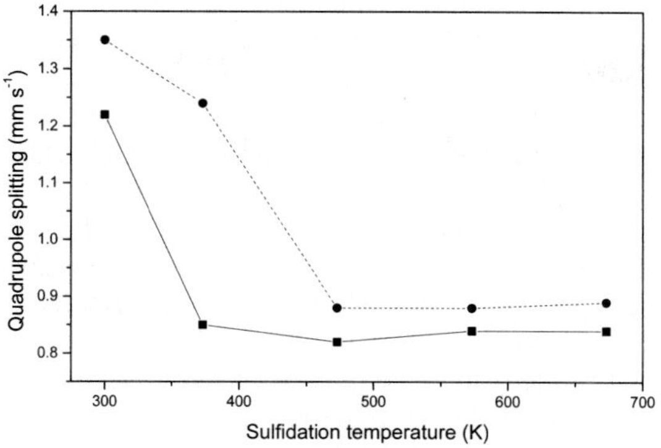

Figure 4.4 *Q.S. values of the Co-sulfide doublet obtained with $Co(2.25)Mo(7)/Al_2O_3$-383 catalyst at 0.1 MPa (●) and at 4 MPa (■) after stepwise sulfidation at the indicated temperatures.*

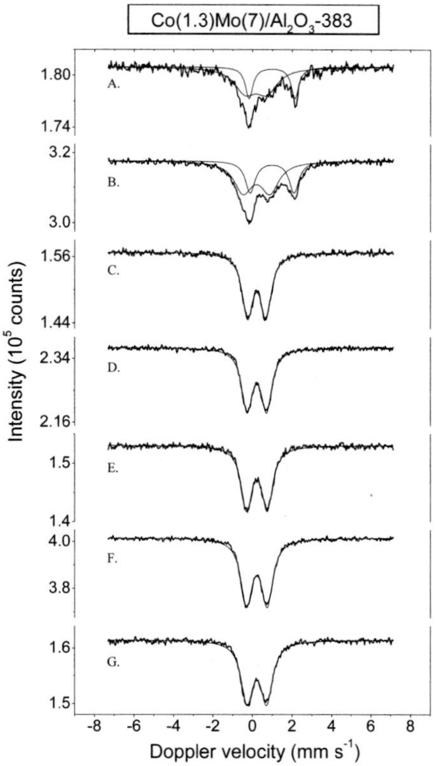

Figure 4.5 *MES spectra obtained at 300 K with $Co(1.3)Mo(7)/Al_2O_3$-383 catalyst after various successive sulfidation steps. A. Fresh catalyst; B. Drying treatment; C. (S, 4 MPa, 473 K); D. (S, 4 MPa, 573 K); E. (S, 4 MPa, 673 K); F. (S, 4 MPa, 723 K); G. (S, 4 MPa, 773 K).*

A further increase of the sulfidation temperature leads to a increasing contribution of such species with lower QS value which points to further segregation concomitant with some agglomeration of the Co-sulfide species.

Figure 4.5 shows the MES spectra obtained for Co(1.3)Mo(7)/Al_2O_3-383 sulfided at 4 MPa. The corresponding fit parameters are presented in Table 4.3. After drying, the catalyst was directly sulfided at 473 K and 4 MPa. The high-spin 2+ spectral contribution disappeared in favor of a doublet with I.S. = 0.21 mm s^{-1} and Q.S. = 0.93 mm s^{-1}. The Q.S. of the remaining doublet increases after sulfidation at 573 K. The resulting doublet has an I.S. value of 0.22 mm s^{-1} and a Q.S. value of 1.05 mm s^{-1}, parameters that are characteristic for the Co-Mo-S phase. After activation at 773 K and 4 MPa, the Q.S. of this latter phase decreases, but in contrast to the results obtained with the high-loading sample no sintering of the Co-sulfide species can be deduced from the MES spectra.

The MES results obtained for Co(1.3)Mo(7)/Al_2O_3-383 sulfided at atmospheric pressure followed by a high-pressure treatment are presented in Figure 4.6 and Table 4.4. Again, the high-spin 2+ contribution completely disappears from the spectrum after treatment at 473 K and 0.1 MPa. The Q.S. value of the remaining spectral contribution increases after sulfidation at 573 K and atmospheric pressure, but no subsequent decrease is observed after atmospheric and high-pressure treatment at 773 K.

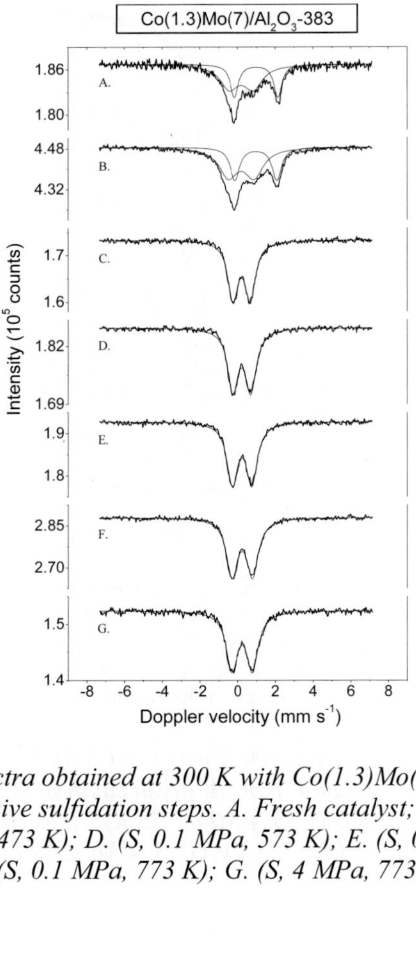

Figure 4.6 *MES spectra obtained at 300 K with Co(1.3)Mo(7)/Al_2O_3-383 catalyst after various successive sulfidation steps. A. Fresh catalyst; B. Drying treatment; C. (S, 0.1 MPa, 473 K); D. (S, 0.1 MPa, 573 K); E. (S, 0.1 MPa, 673 K); F. (S, 0.1 MPa, 773 K); G. (S, 4 MPa, 773 K).*

Table 4.3 *MES parameters of Co(1.3)Mo(7)/Al_2O_3-383 catalyst after sulfidation treatment*

T_s K	P MPa	IS mm s^{-1}	QS mm s^{-1}	Γ mm s^{-1}	A %	IS mm s^{-1}	QS mm s^{-1}	Γ mm s^{-1}	A %	IS mm s^{-1}	QS mm s^{-1}	Γ mm s^{-1}	A %
		"Co – oxide"				"Co – sulfide"				"High-spin 2+"			
Fresh		0.16	1.09	1.32	67					0.98	2.32	0.46	33
H_2/393		0.19	1.35	1.11	63					0.98	2.22	0.61	37
473	4					0.21	0.93	0.66	100				
573	4					0.22	0.98	0.66	100				
673	4					0.22	1.05	0.66	100				
723	4					0.22	1.05	0.69	100				
773	4					0.22	0.99	0.72	100				

Table 4.4 *MES parameters of Co(1.3)Mo(7)/Al_2O_3-383 catalyst after sulfidation treatment*

T_s K	P MPa	IS[a] mm s^{-1}	QS[b] mm s^{-1}	Γ[c] mm s^{-1}	A[d] %	IS mm s^{-1}	QS mm s^{-1}	Γ mm s^{-1}	A %	IS mm s^{-1}	QS mm s^{-1}	Γ mm s^{-1}	A %
		"Co – oxide"				"Co – sulfide"				"High-spin 2+"			
Fresh		0.20	1.28	1.19	61					0.99	2.31	0.51	39
H_2/393		0.19	1.37	1.20	62					0.97	2.23	0.60	38
473	0.1					0.21	0.93	0.66	100				
573	0.1					0.23	0.96	0.68	100				
673	0.1					0.23	1.04	0.69	100				
773	0.1					0.24	1.10	0.73	100				
773	4					0.23	1.10	0.73	100				

The evolution of the Q.S. values for the Co(1.3)Mo(7)/Al_2O_3-383 catalyst as a function of temperature during sulfidation at atmospheric and high-pressure conditions is presented in Figure 4.7.

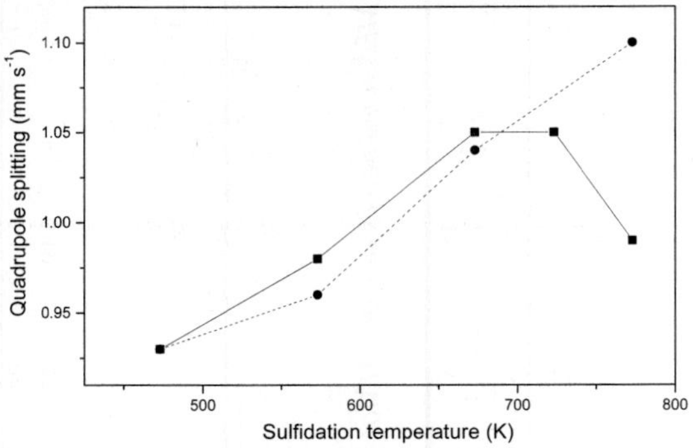

Figure 4.7 *Q.S. values of the Co-sulfide doublet obtained with Co(1.3)Mo(7)/Al_2O_3-383 catalyst at 0.1 MPa (●) and at 4 MPa (■) after stepwise sulfidation at the indicated temperatures.*

4.3.2 Mo K edge EXAFS

Mo K edge EXAFS fit parameters of Co(2.25)Mo(7)/Al_2O_3-383 catalyst sulfided at different temperatures and pressures (0.1 and 4 MPa) are collected in Table 4.5.

Table 4.5 *Mo K-edge EXAFS parameters of Co(2.25)Mo(7)/ Al_2O_3-383 catalyst after different sulfidation treatments*

Treatment	N -	R Å	$\Delta\sigma^2$ Å^2 $\cdot 10^{-3}$	ΔE_0 eV	N -	R Å	$\Delta\sigma^2$ Å^2 $\cdot 10^{-3}$	ΔE_0 eV	N -	R Å	$\Delta\sigma^2$ Å^2 $\cdot 10^{-3}$	ΔE_0 eV
	Mo-S				Mo-Mo				Mo-Co			
S, 4 MPa, 473 K	3.6	2.46	3.8	1.2	1.4	3.24	8.6	9.8				
S, 4 MPa, 573 K	5.5	2.41	1.9	0.3	3.1	3.16	4.4	2.0	0.9	2.75	9.8	9.3
S, 4 MPa, 673 K	5.6	2.41	1.3	1.6	3.9	3.16	7.8	2.5	0.9	2.76	9.9	4.7
S, 0.1 MPa, 473 K	4.3	2.41	7.8	-0.5	0.9	3.10	5.4	8.9				
S, 0.1 MPa, 573 K	4.4	2.41	4.1	3.5	1.7	3.15	6.8	4.9				
S, 0.1 MPa, 673 K	5.1	2.41	2.2	1.0	2.7	3.15	4.2	3.2	0.7	2.75	9.8	9.7
S, 0.1 MPa + S, 4 MPa, 673 K	5.4	2.41	2.3	1.1	2.9	3.15	4.3	3.0	0.7	2.75	9.5	9.9

Upon sulfidation at 4 MPa, the Mo-S and Mo-Mo interatomic distances are in close agreement with the structural parameters of bulk MoS_2 [23].

The coordination numbers are lower than those of bulk MoS_2 ($N_{Mo\text{-}S}$ = 6, $N_{Mo\text{-}Mo}$ = 6), because of the limited size of the MoS_2 particles. In addition to these contributions, after sulfidation at 573 K and 4 MPa, a Mo-Co contribution is observed at a distance of about 2.75 Å. The $N_{Mo\text{-}Co}$ remains constant following further sulfidation at 673 K and 4 MPa.

When the sulfidation is carried out at atmospheric pressure, a very similar behavior is observed. The presence of MoS_2-like species is again evidenced by Mo-Mo and Mo-S coordination distances close to 2.41 A and 3.16, respectively, at all temperatures. In this case, the Mo-Co contribution is observed only after activation at 673 K. After the subsequent sulfidation at 4 MPa and 673 K, only a small increase in the $N_{Mo\text{-}Mo}$ value is found. The resulting $N_{Mo\text{-}S}$ coordination number is somewhat smaller compared to that of the catalyst sulfided directly at high-pressure.

4.3.3 Co EXAFS

The Co K-edge EXAFS fit parameters of the sulfided Co(2.25)Mo(7)/Al_2O_3-383 catalyst are listed in Table 4.6. After treatment at 4 MPa, the spectra were best fitted with a Co-S contribution at a distance of 2.22 Å and two Co-Co contributions at 2.6 Å and at 3.41 Å, distances that are slightly different from those corresponding to bulk Co_9S_8 (2.51 Å and 3.51 Å, respectively) [14]. A Co-Mo contribution at 2.8 Å was included in the spectra measured after treatment at 573 K and 673 K.

Upon activation of the same catalyst at atmospheric pressure, the Co-Mo contribution is observed only upon sulfidation at 673 K. Moreover, $N_{Co\text{-}Mo}$ increases from 0.5 to 1.0 after the subsequent sulfidation at 673 K and 4 MPa. The Co-S coordination number is similar for the samples sulfided at 673 K and different pressures, while a somewhat larger $N_{Co\text{-}Co(1)}$ is obtained with the sample sulfided directly at high-pressure.

Similar results were obtained with the Co(1.3)Mo(7)/Al_2O_3-383 catalyst sulfided at atmospheric pressure (Table 4.7). Again, the Co-Mo contribution is observed only after treatment at 673 K, with $N_{Co\text{-}Mo}$ being twice as high as observed with the high-loading sample.

4.3.4 TEM

Figure 4.8 shows representative TEM micrographs of the Co(2.25)Mo(7)/Al_2O_3-383 catalyst sulfided at 673 K and different pressures. The mean slab lengths and the stacking degrees are summarized in Table 4.8. These results were obtained by evaluating at least 300 MoS_2 slabs per sample in at least three representative images. The average slab length and the degree of stacking are higher for the sample sulfided at 673 K and 4 MPa compared to the sample sulfided at atmospheric pressure. The slabs are also better defined, indicating increased crystallinity after high-pressure treatment. Only a small difference in the slab length and stacking degree is observed when increasing the sulfidation pressure from 0.1 to 4 MPa.

Table 4.6 *Co K-edge EXAFS parameters of Co(2.25)Mo(7)/ Al_2O_3-383catalyst after different sulfidation treatments*

Treatment	N -	R Å	$\Delta\sigma^2$ $Å^2$ $\cdot 10^{-3}$	ΔE_0 eV	N -	R Å	$\Delta\sigma^2$ $Å^2$ $\cdot 10^{-3}$	ΔE_0 eV	N -	R Å	$\Delta\sigma^2$ $Å^2$ $\cdot 10^{-3}$	ΔE_0 eV	N -	R Å	$\Delta\sigma^2$ $Å^2$ $\cdot 10^{-3}$	ΔE_0 eV
	Co-S				Co-Co (1)				Co-Mo				Co-Co (2)			
S, 4 MPa, 300 K	5.9	2.23	3.1	0.8	0.9	2.50	5.0	5.6					1.4	3.34	9.2	-4.7
S, 4 MPa, 373 K	5.8	2.23	3.8	0.3	0.9	2.50	4.2	5.4					1.3	3.34	9.8	-1.8
S, 4 MPa, 473 K	5.6	2.23	3.8	0.5	0.9	2.52	5.1	-0.1					0.8	3.34	3.6	-1.2
S, 4 MPa, 573 K	5.6	2.22	3.6	-0.3	0.5	2.56	9.9	-9.8	0.7	2.80	9.8	-7.4	0.4	3.41	2.4	-7.8
S, 4 MPa, 673 K	5.7	2.22	4.4	0.6	1.1	2.60	9.9	-9.7	0.7	2.81	8.4	-9.4	0.7	3.41	9.7	-1.8
S, 0.1 MPa, 373 K	6.0	2.23	2.8	1.7	1.2	2.50	7.5	0.1					1.4	3.21	9.2	5.2
S, 0.1 MPa, 473 K	5.8	2.23	3.2	-1.6	1.0	2.56	4.4	-0.5					1.0	3.23	7.0	9.9
S, 0.1 MPa, 573 K	5.6	2.23	3.8	-0.8	1.2	2.59	9.7	-8.0					1.2	3.26	5.9	9.8
S, 0.1 MPa, 673 K	5.7	2.22	4.4	0	0.8	2.59	9.7	-8.4	0.5	2.80	4.1	-3.6	0.7	3.41	9.8	-2.5
S, 0.1 MPa + S, 4 MPa, 673 K	5.8	2.23	4.9	-0.7	0.7	2.60	9.9	-9.5	1.0	2.81	9.8	-6.9	0.9	3.41	9.5	-4.2

Table 4.7 *Co K-edge EXAFS parameters of Co(1.3)Mo(7)/ Al_2O_3-383catalyst after different sulfidation treatments*

Treatment	N -	R Å	$\Delta\sigma^2$ $Å^2$ $\cdot 10^{-3}$	ΔE_0 eV	N -	R Å	$\Delta\sigma^2$ $Å^2$ $\cdot 10^{-3}$	ΔE_0 eV	N -	R Å	$\Delta\sigma^2$ $Å^2$ $\cdot 10^{-3}$	ΔE_0 eV	N -	R Å	$\Delta\sigma^2$ $Å^2$ $\cdot 10^{-3}$	ΔE_0 eV
	Co-S				Co-Co (1)				Co-Mo				Co-Co (2)			
S, 0.1 MPa, 473 K	5.4	2.23	3.9	-0.4	1.2	2.58	5.6	-5.1					1.2	3.28	9.8	9.2
S, 0.1 MPa, 573 K	5.3	2.23	3.7	-0.8	1.3	2.57	5.8	-2.5					1.3	3.28	9.5	9.3
S, 0.1 MPa, 673 K	5.6	2.23	4.3	-1.2	0.9	2.57	9.2	-1.4	1.1	2.80	9.5	-5.9	0.5	3.41	9.3	-3.5

Figure 4.8 *TEM images of Co(2.25)Mo(7)/Al_2O_3-383 catalyst after various sulfidation treatments. Left: (S, 4 MPa, 673 K); middle: (S, 0.1 MPa, 673 K); right: (S, 0.1 MPa, 673 K + S, 4 MPa, 673 K).*

Table 4.8 *Slab length and stacking degree of Co(2.25)Mo(7)/ Al_2O_3-383 catalyst after different sulfidation treatments*

Treatment	*Average slab length (nm)*	*Average stacking*
S, 4 MPa, 673 K	4.2	1.8
S, 0.1 MPa, 673 K	3.0	1.3
S, 0.1 MPa, 673 K + S, 4 MPa, 673 K	3.2	1.4

4.3.5 Dibenzothiophene HDS

The DBT HDS activities of the Co(2.25)Mo(7)/Al_2O_3-383 catalyst activated at different pressures and temperatures are shown in Table 4.9. The first-order rate constant (k_{HDS}), measured after 8 hours at 3 MPa and 573 K, is expressed per kg of Mo. The apparent activation energies (E_{act}) and the pre-exponential factors (ν_{pre}) were determined in the 533-573 K temperature interval.

Table 4.9 *DBT HDS activities, apparent activation energies and pre-exponential factors of Co(2.25)Mo(7)/ Al_2O_3-383 catalyst after different sulfidation treatments*

Treatment	k_{HDS} (mol DBT·$Kg_{Mo}^{-1}\cdot s^{-1}$)·10^{-2}	E_{act} (kJ·mol^{-1})	ν_{pre} (mol·$kg^{-1}\cdot s^{-1}$)
S, 4 MPa, 673 K	2.1	121	$2.2\cdot10^{9}$
S, 0.1 MPa, 673 K	1.1	103	$2.8\cdot10^{7}$
S, 0.1 MPa, 673 K + S, 4 MPa, 673 K	1.8	91	$3.3\cdot10^{6}$
S, 4 MPa, 723 K	1.6	105	$5.6\cdot10^{7}$

The activity is highest for the sample directly sulfided at 4 MPa and 673 K. Increasing the sulfidation temperature to 723 K at the same pressure leads to a substantial decrease of the activity. Sulfidation at atmospheric pressure leads to a two times lower activity. Subsequent sulfidation at high pressure increases the activity substantially. The apparent activation energy and the pre-exponential factor of the catalyst sulfided at 673 K and 4 MPa are considerably higher than for the other samples. After treatment at 4 MPa and 723 K a small decrease in the activity is observed.

The activity trends for the catalyst with a lower Co content are very similar (Table 4.10). The overall activities are slightly lower than for the catalysts with a higher Co content.

Table 4.10 *DBT HDS activities, apparent activation energies and pre-exponential factors of Co(1.3)Mo(7)/ Al_2O_3-383 catalyst after different sulfidation treatments*

Treatment	k_{HDS} *(mol DBT·Kg* $_{Mo}^{-1}$·s^{-1}) ·10^{-2}	E_{act} *(kJ·mol^{-1})*	v_{pre} *(mol·kg^{-1}·s^{-1})*
S, 4 MPa, 673 K	1.8	110	$1.9·10^8$
S, 0.1 MPa, 673 K	0.9	86	$5.7·10^5$
S, 0.1 MPa, 673 K + S, 4 MPa, 673 K	1.3	93	$3.6·10^6$

4.4 Discussion

The MES spectra of all fresh samples consist of one quadrupole doublet assigned to a low-spin 2+ or a 3+ phase and a second contribution consistent with Co^{2+} in the high-spin state [22]. The latter is associated with Co ions interacting strongly with the γ-Al_2O_3 support. An increase in the spectral contribution of the high-spin 2+ phase is observed with the Co(2.25)Mo(7)/Al_2O_3-383 catalyst after the drying treatment and can be explained by removal of water associated with the high-spin species in the fresh catalyst, causing an increased interaction of this species with the support [25]. This effect is less significant for the Co(1.3)Mo(7)/Al_2O_3-383 catalyst, possibly because of the formation of smaller intermediate Co species having less affinity for water.

After room temperature exposure of Co(2.25)Mo(7)/Al_2O_3 to the H_2S/H_2 mixture at atmospheric pressure, an additional increase of the high-spin 2+ doublet is noticed as a result of the formation of an intermediate Co-sulfate type phase [16]. The presence of a second doublet with an I.S. value of 0.23 mm s^{-1} and a Q.S. value of 1.34 mm s^{-1}, assigned to a Co-sulfide phase, indicates that the sulfidation of the catalysts starts already at room temperature. At room temperature already 43% of the Co species is in this form. This contribution increases to 51% when the sulfidation pressure is increased. The Q.S. of this doublet decreases to 0.82 mm s^{-1} after further high-pressure sulfidation at 373 K and 473 K. This points to an increase in the particle size of these Co-sulfide species. Concomitantly, the high-spin 2+ spectral contribution gradually decreases with increasing sulfidation pressure and temperature and completely disappears from the spectra upon treatment at 473 K and 4 MPa. Sulfidation of a calcined Co(2.25)Mo(7)/Al_2O_3 catalyst (Chapter 3) showed that a sulfidation temperature of 673 K was required to sulfide all the Co promoter ions.

In contrast to this, when the catalyst is not calcined upon introduction of the promoter ions in the second step, treatment at 473 K suffices to sulfide all the Co atoms.

Upon high-pressure treatment at 573 K and 673 K only a small increase in the Q.S. value is observed. This points to a reduced redispersion of the Co-sulfide particles over the edges of the MoS_2 crystallites. This can be understood in terms of a weak Co-Mo interaction leading to the partial formation of small CoS_x particles (highly dispersed Co_9S_8 type structures). The rest of the Co atoms are located on edge positions of the MoS_2 in Co-Mo-S structures. The formation of highly dispersed Co_9S_8-like species having MES parameters that resemble the Co-Mo-S parameters was also observed with carbon-supported catalysts and uncalcined alumina-supported catalysts, sulfided under atmospheric pressure conditions [16,25].

The MES spectrum of the Co(2.25)Mo(7)/Al_2O_3-383 catalyst sulfided at 673 K and 4 MPa (Figure 4.1, H) was fitted with only one quadrupole doublet. However, the Q.S. value of this doublet is smaller compared to that of the calcined sample measured after the same treatment (Chapter 3). This might be indicative of the presence of a broad range of CoS_x species differing in size and ordering. Therefore, this doublet may be alternatively fitted with a distribution in Q.S. values accounting for the presence of such CoS_x particles [25]. Although this has no direct physical meaning, a simplified approach is to fit the spectrum with two doublets (Figure 4.9), the resulted fit parameters indicating that about 60 % of the Co atoms are located in CoS_x structures and the rest are forming the Co-Mo-S species.

That this approach is reasonable is underpinned by the segregation of about 56 % of the Co atoms upon treatment at 723 K and 4 MPa (Figure 4.1, I). The Q.S. value of 0.44 mm s^{-1} of the new-formed doublet is larger than the Q.S. value of 0.26 mm s^{-1} for crystalline Co_9S_8 [2], and this doublet is assigned to a dispersed Co_9S_8-like species. Upon the more severe sulfidation treatment, the initially present CoS_x structures sinter and the resulting Co-sulfide species are larger and approach crystalline Co_9S_8 in size and in Mössbauer parameters [16].

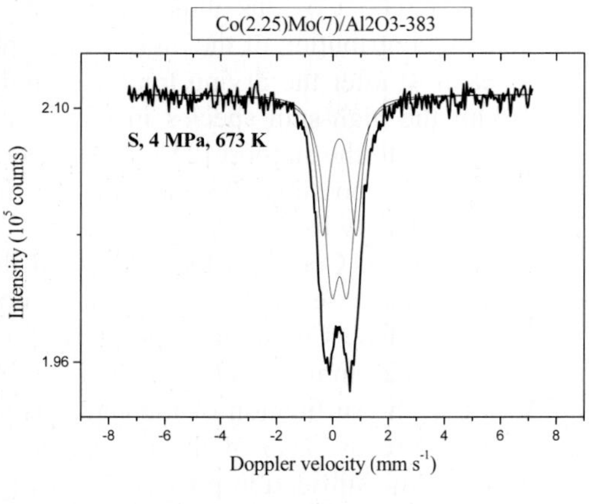

Figure 4.9 *Alternative fit for the MES spectrum obtained at 300 K with Co(2.25)Mo(7)/Al_2O_3-383 catalyst after sulfidation at 673 K and 4 MPa.*

The Co(2.25)Mo(7)/Al_2O_3-383 catalyst activated first under atmospheric pressure conditions shows sulfidation behavior similar to the same sample treated directly at high-pressure (see Figure 4.4). However, the high-spin 2+ contribution is still large after sulfidation at 373 K and does not completely disappear from the spectrum upon treatment at 473 K. The sulfidation rate is lower at 0.1 MPa, the somewhat larger Q.S. values of the second spectral contribution pointing to the formation of smaller Co-sulfide species. No increase in the Q.S. is observed after treatment at 573 K and atmospheric pressure, a good indication that again a large amount of CoS_x particles is formed.

The Q.S. values of 0.89 mm s^{-1} and 0.86 mm s^{-1} of the dominant doublet, measured after sulfidation at 673 K and 0.1 MPa and after the subsequent increase in pressure up to 4 MPa, are also smaller compared to that of the calcined catalyst (Chapter 3). In the same way as for the sample sulfided at high-pressure (Figure 4.9), it can be shown that about 50-60 % of the Co atoms are present in the CoS_x structures. Upon treatment at 723 K and 4 MPa (Figure 4.3, F), these CoS_x particles sinter and grow in size after subsequent sulfidation at 773 K, approaching crystalline Co_9S_8.

The Co(1.3)Mo(7)/Al_2O_3-383 catalyst is also completely sulfided after treatment at 473 K (at 0.1 and 4 MPa). The weaker Co-support interaction in uncalcined catalysts allows the participation of all Co atoms in Co-sulfide species at intermediate temperatures. A strong increase in the Q.S. value of the main doublet is observed during treatment at 573 K and 673 K, under both atmospheric and high-pressure conditions (Figure 4.7). This indicates a better redispersion of the Co-sulfide particles over the edges of newly formed MoS_2 crystallites developed at temperatures of about 573 K [26].

The doublets obtained after sulfidation at 673 K, at both 0.1 and 4 MPa, have Q.S. values of 1.04 mm s^{-1} and 1.05 mm s^{-1}, parameters that are characteristic for the presence of the Co-Mo-S phase. These values are higher than those obtained for the Co-Mo-S structures in calcined catalysts having larger Co/Mo ratio [27]. This is because the Q.S. values of the Co-Mo-S doublets are generally higher for a smaller Co/Mo ratio [14]. A small decrease in the Q.S. value of the Co-Mo-S doublet is observed at 773 K with the sample treated directly under high-pressure conditions. This indicates that sintering of the Co-sulfide species will take place faster in the catalyst sulfided at 4 MPa, probably as a result of the formation of larger MoS_2 crystallites under these conditions. A contribution of segregated CoS_x species is not observed in the spectra of sulfided Co(1.3)Mo(7)/Al_2O_3-383 catalyst. This coheres with the notion that Co_9S_8 segregation is shifted to higher temperatures with decreasing Co content [28].

The Co K-edge EXAFS fit results for Co(2.25)Mo(7)/Al_2O_3-383 (Table 4.6) show that the Co atoms are sulfided already after exposure to the H_2S/H_2 gas mixture at room temperature under atmospheric and high-pressure conditions. After treatments up to 373 K at both 0.1 MPa and 4 MPa, the resulted Co-S (2.23 Å) and Co-Co (1) (2.50 Å) coordination distances are similar to the values previously reported for Co_9S_8 [14, 29]. The long Co-Co(2) coordination distance (3.34 Å) – close to the corresponding distance in Co_9S_8 – and the somewhat high Co-S and Co-Co (1) coordination numbers obtained upon high-pressure activation, indicate the formation of larger Co-sulfide species as a result of the higher rate of Co sulfidation under these conditions. A Co-S coordination number of about 6 is measured after treatment at all intermediate temperatures under atmospheric and high-pressure conditions, indicating that no large amounts of Co_9S_8 are formed up to 673 K. This agrees with the MES results that only clearly show the presence of a Co_9S_8-like doublet after activation at 723 K (Figure 4.1, I and Figure 4.3, F). Also the relatively small Co-Co coordination numbers observed support such conclusion. Similar results obtained with CoMo/C [16] and NiMo/SiO_2 catalysts [30] suggested the formation of sulfided species with very disordered Co-S structures and coordinations different from those of bulk crystalline cobalt sulfide systems. These special structures are formed as soon as the catalysts are sulfided and after treatment at 573 and 673 K only part of the Co atoms will participate in the Co-Mo-S species. A partial redispersion of the Co-sulfide particles at the temperature of MoS_2 crystallites formation can be deduced from the increase in the Co-Co(1) atomic distance at 573 K. A similar redispersion model has been proposed for NiS_x particles in NiW catalysts [31].

The EXAFS data analysis of the Co(2.25)Mo(7)/Al_2O_3-383 sample sulfided at 573 K and 4 MPa reveals the presence of a Co-Mo contribution at 2.8 Å, in good agreement with the previously reported values [16,32]. This is in line with the conclusion that Co-Mo-S structures are formed at this temperature. For the sample treated under atmospheric pressure conditions, a smaller Co-Mo coordination shell is observed only after sulfidation at 673 K. The higher Co-Mo coordination numbers obtained with the catalysts activated at 4 MPa point in the direction of Type II Co-Mo-S phase formation upon increasing sulfidation pressure. The Co species become better defined upon sulfidation at higher pressure. Because of the partial formation of CoS_x species, the observed Co-Mo coordination numbers are lower compared with those obtained with the similar calcined sample (Chapter 3). Also the Co(1.3)Mo(7)/Al_2O_3-383 catalyst sulfided at 0.1 MPa (Table 4.7) shows the presence of a Co-Mo contribution only after treatment at 673 K. This indicates a lower degree of sulfidation and is consistent with the observation that Type I Co-Mo-S species are formed under atmospheric pressure conditions.

The Mo K-edge EXAFS fit results obtained for Co(2.25)Mo(7)/Al_2O_3-383 sulfided under atmospheric and high-pressure conditions (Table 4.5), show the presence of MoS_2-like species at all intermediate temperatures. After treatment at 573 K and 673 K, the Mo-S and Mo-Mo interatomic distances are in agreement with the structural parameters of MoS_2 crystallites [23]. A relatively low Mo-S coordination number ($N_{Mo-S} = 5.1$) is obtained after treatment at 673 K and 0.1 MPa, indicating that the MoS_2 edges are not completely saturated with sulfur, probably due to the presence of some Mo-O-Al linkages of Mo atoms with the alumina support or at least incomplete sulfur coordination [33]. This agrees with the definition of the Type I Co-Mo-S phase, which was also formed in the similar calcined catalyst under the same sulfidation conditions (Chapter 3).

The Mo-Co contributions are observed starting with the treatment at 573 K for the sample sulfided directly at 4 MPa, and only after sulfidation at 673 K for the catalyst treated under atmospheric pressure conditions. These results agree well with the Co EXAFS fit results.

Higher Mo-S and Mo-Mo coordination numbers were measured after sulfidation at 673 K and 4 MPa, pointing to an accelerated Mo sulfidation under high-pressure conditions. The EXAFS data indicate a better saturation of the MoS_2 particles with sulfur and the formation of somewhat larger or better-defined crystallites. This also holds for the atmospheric pressure sulfided sample when subsequent to the atmospheric pressure sulfidation the sulfidation pressure is increased. All these results suggest the formation of the Type II Co-Mo-S phase upon activation at 4 MPa, and likely only a partial transition of the active Co-Mo-S structures to Type II species for the catalyst treated first at 0.1 MPa.

The TEM micrographs of Co(2.25)Mo(7)/Al_2O_3-383 sulfided at 673 K in Figure 4.8 show higher average slab length and stacking degree for the catalyst sulfided directly at high pressure. The more sintered character of the MoS_2 phase upon direct high-pressure sulfidation is caused by the accelerated sulfidation leading to decreased interaction with the subjacent support and, as suggested earlier [12,34] to sintering due to slow removal of locally generated heat. These changes go together with increased crystallinity and can also be related to a higher mobility of the molybdenum species on the catalyst surface.

The degree of stacking and the average slab length increase slightly also for the atmospheric pressure sulfided sample after the subsequent activation at 4 MPa. For both catalysts treated at elevated pressure, better-defined slabs are observed which indicates increased crystallinity of the MoS_2 phase. This confirms to some extent the conclusion that Type II Co-Mo-S species are formed after the 4 MPa treatment.

The DBT HDS activities of the Co(2.25)Mo(7)/Al_2O_3-383 catalyst sulfided at 673 K and 4 MPa, either directly or after atmospheric sulfidation, are approximately two times higher than those of the catalyst sulfided at 0.1 MPa. This ratio is in good agreement with the relative activities of the Type II and I Co-Mo-S species observed in gas phase thiophene HDS [35,36]. The activity of the sample sulfided first under atmospheric pressure and subsequently at 4 MPa is somewhat smaller compared to that of the catalyst activated directly under high-pressure conditions, possibly because of an incomplete formation of Type II structures.

The DBT HDS activities of the Co(1.3)Mo(7)/Al_2O_3-383 catalysts are only marginally lower than for the Co(2.25)Mo(7)/Al2O3-383 catalyst. This appears to be in accordance with the proposed limited contribution of about 40% of the Co atoms in the latter catalyst to Co-Mo-S structures. The remainder are CoS_x species that contribute little to the catalytic performance. The activity of the Co(2.25)Mo(7)/Al_2O_3-383 catalyst obtained after activation at 723 K and 4 MPa is lower compared to that of the sample sulfided at 673 K, probably because of the partial blocking of the active sites by the larger Co_9S_8-like species.

The apparent activation energies are higher for the catalysts activated under high-pressure conditions than for the atmospheric pressure sulfided samples. This difference may indicate a change in the interaction energy between the catalytic surface and the reactant [37] and could relate to the type of Co-Mo-S species present. Such changes in interaction energy are also reflected in the pre-exponential factor [38].

4.5 Conclusions

The sulfidation of uncalcined $CoMo/Al_2O_3$ catalysts under conditions relevant to industrial practice was studied by MES, EXAFS, TEM and DBT HDS. The absence of the calcination step after loading of the calcined Mo/Al_2O_3 catalyst with Co leads to a decreased interaction of the promoter ions with the Al_2O_3 support. As a consequence, all promoter atoms are completely sulfided at lower temperatures than when calcination would have been applied. This also leads to a significant formation of small CoS_x particles or highly dispersed Co_9S_8-type species with a low activity when a relatively high Co/Mo ratio of 0.52 is used. Although the CoS_x structures can have MES parameters that resemble those of the Co-Mo-S phase, MES can still distinguish between the different Co-sulfides present, since smaller Q.S. values are clearly indicating the presence of CoS_x species involving variable number of Co atoms.

The stepwise sulfidation approach points to somewhat higher sulfidation rates of Co and Mo during high-pressure activation. The more aggregated Co-sulfide structures formed at intermediate temperatures show a reduced redispersion of the cobalt species for the

$Co(2.25)Mo(7)/Al_2O_3$-383 catalyst, once MoS_2 is formed. This agrees with a larger contribution of CoS_x particles in such a catalyst.

The final state of the catalysts after 0.1 MPa or 4 MPa sulfidation is very similar in terms of Q.S. of the Co-sulfide species. Yet, the high-pressure sulfided samples have a much higher HDS activity than the samples activated at 0.1 MPa. The difference is due to formation of Type II Co-Mo-S phase at elevated pressures. The higher Co-Mo coordination numbers and the more crystalline MoS_2 slabs obtained with the catalysts treated also at 4 MPa give support to the idea of formation of Type II Co-Mo-S structures. The activities of the samples with Co/Mo ratios of 0.30 and 0.52 are very similar which further underpins the finding that for the catalyst with the higher promoter content a substantial part of the Co has ended up in segregated CoS_x structures.

References

1. B.S. Clausen, S. Mørup, H. Topsøe, R. Candia, J. Phys. 37 (1976) 249.
2. H. Topsøe, B.S. Clausen, R. Candia, C. Wivel, S. Mørup, J. Catal. 68 (1981) 433.
3. C. Wivel, R. Candia, B.S. Clausen, S. Mørup, H. Topsøe, J. Catal. 68 (1981) 453.
4. R. Candia, O. Sorensen, J. Villadsen, N.-Y. Topsøe, B.S. Clausen, H. Topsøe, Bull. Soc. Chim. Belg. 93 (1984) 763.
5. C. Wivel, R. Candia, B.S. Clausen, S. Mørup and H. Topsøe, J. Catal. 87 (1984) 497.
6. J.T. Richardson, Ind. Eng. Chem. Fund. 3 (1964) 154.
7. I.E. Wachs, Catal. Today 27 (1996) 437.
8. A.F.H. Sanders, A.M. de Jong, V.H.J. de Beer, J.A.R. van Veen, J.W. Niemantsverdriet, Appl. Surf. Sci. 144-145 (1999) 380.
9. Y.S. Al-Zeghayer, B.Y. Jibril, Appl. Catal. A 292 (2005) 287.
10. B. Scheffer, N.J.J. Dekker, P.J. Mangnus, J.A. Moulijn, J. Catal. 121 (1990) 31.
11. Y. Okamoto, S. Ishihara, M. Kawano, M. Satoh and T. Kubota, J. Catal. 217 (2003) 12.
12. P.J. Kooyman, J.G. Buglass, H.R. Reinhoudt, A.D. van Langeveld, E.J.M. Hensen, H.W. Zandbergen and J.A.R. van Veen, J. Phys. Chem. B 106 (2002) 11795.
13. S.M.A.M. Bouwens, R. Prins, V.H.J. de Beer, D.C. Koningsberger, J. Phys. Chem. 94 (1990) 3711.
14. S.M.A.M. Bouwens, J.A.R. van Veen, D.C. Koningsberger, V.H.J. de Beer, R. Prins, J. Phys. Chem. 95 (1991) 123.
15. T. Shido, R. Prins, J. Phys. Chem. 102 (1998) 8426.
16. M.W.J. Crajé, S.P.A. Louwers, V.H.J. de Beer, R. Prins, A.M. van der Kraan, J. Phys. Chem. 96 (1992) 5445.
17. M.W.J. Crajé, V.H.J. de Beer, J.A.R. van Veen and A.M. van der Kraan, J. Catal. 143 (1993) 601.
18. S. Houssenbay, S. Kasztelan, H. Toulhoat, J.P. Bonnelle, J. Grimblot, J. Phys. Chem. 93 (1989) 7176.
19. J.V. Lauritsen, S. Helveg, E. Lægsgaard, I. Stensgaard, B.S. Clausen, H. Topsøe, F. Besenbacher, J. Catal. 197 (2001) 1.
20. M.W.J. Crajé, Ph.D. Thesis, Delft University of Technology, Delft, 1992, ISBN 90-73861-08-X.

21. M.W.J. Crajé, A. M. van der Kraan, J. van de Loosdrecht and P. J. van Berge, Catal. Today 71 (2002) 369.
22. J.G. Steven, L. Zhe, H. Pollak, V.E. Stevens, R.H. White and J.L. Gibson, Mössbauer Handbook Minerals, Mössbauer Effect Data Center, University of North Carolina, Asheville, 1983.
23. T.G. Parham, R.P. Merrill, J. Catal. 85 (1984) 295.
24. W.L.T.M. Ramselaar, M.W.J. Crajé, R.H. Hadders, E. Gerkema, V.H.J. de Beer, A.M. van der Kraan, Appl. Catal. 65 (1990) 69.
25. M.W.J. Crajé, Ph.D. Thesis, Delft University of Technology, Delft, 1992, ISBN 90-73861-08-X.
26. M. de Boer, A.J. van Dillen, D.C. Koningsberger, J.W. Geus, Jpn. J. Appl. Phys. 32 (1992) 460.
27. A.I. Dugulan, M.W.J. Crajé and G.J. Kearley, J. Catal. 222 (2004) 281.
28. R. Candia, B.S. Clausen, H. Topsøe, Effect of Sulfiding Temperature on Activity and Structure of Co-Mo/Al_2O_3 Catalysts, Proc. 9th Iberoamerican Symp. Catal., 16-21 July 1984, Lisbon, Portugal, 1984, p. 211.
29. W. Niemann, B.S. Clausen, H. Topsøe, Catal. Lett. 4 (1990) 355.
30. L. Medici, R. Prins, J. Catal. 163 (1996) 38.
31. E.J.M. Hensen, Y. van der Meer, J.A.R. van Veen and J.W. Niemantsverdriet, Appl. Catal. A, 322 (2007) 16.
32. S.M.A.M. Bouwens, D.C. Koningsberger, V.H.J. de Beer, S.P.A. Louwers, R. Prins, Catal. Lett. 5 (1990) 273.
33. D. Nicosia, R. Prins, J. Catal. 231 (2005) 259.
34. Y. van der Meer, Ph.D. Thesis, Delft University of Technology, Delft, 2001, ISBN 90-407-2230-7.
35. S.P. Cramer, K.S. Liang, A.J. Jacobson, C.H. Chang, R.R. Chianelli, Inorg. Chem. 23 (1984) 215.
36. J.A.R. van Veen, H.A. Colijn, P.A.J.M. Hendriks, A.J. van Welsenes, Fuel. Proc. Technol. 35 (1993) 137.
37. E. J. M. Hensen, M. J. Vissenberg, V. H. J. de Beer, J. A. R. van Veen, R. A. van Santen, J. Catal. 163 (1996) 429.
38. E. J. M. Hensen, Ph.D. Thesis, Eindhoven University of Technology, Eindhoven, 2000, ISBN 90-386-2871-4.

Chapter 5

The effect of high-pressure sulfidation on the properties of CoMo/C catalysts

Abstract

The sulfidation of CoMo/C catalysts under conditions relevant to industrial practice was studied by Mössbauer emission spectroscopy (MES), extended X-ray absorption fine structure (EXAFS), and dibenzothiophene hydrodesulfurization (HDS) activity measurements. The weak interaction of the promoter atoms with the support and also the weak Co-Mo interaction leads to a significant formation of more aggregated Co-sulfide structures after sulfidation at 0.1 MPa. MoS_2 hinders the complete sintering of the CoS_x structures under atmospheric pressure conditions, but upon more severe sulfidation at 4 MPa these CoS_x species sinter to Co_9S_8-type structures. The higher Co-Mo coordination numbers and the accelerated Mo sulfidation observed with the catalysts treated at 4 MPa give support to the idea of formation of Type II Co-Mo-S structures. The atmospheric pressure sulfidation renders the presence of the active species (highly dispersed Type II Co-Mo-S) constrained in the micropores of the carbon support, resulting in considerably lower DBT HDS activity as compared to the samples activated at 4 MPa.

5.1 Introduction

The "zero-sulphur" fuels (defined as having maximum sulphur content of 10 ppm) become mandatory in 2009 [1]. The continuing tightening of the fuel quality standards results in an increased need for more active hydrodesulfurization (HDS) catalysts.

Extensive research has been carried out on Co(Ni)Mo sulfide hydrotreating catalysts, but the exact nature of the active phase or its mode of operation are still subject of debate. Different structural models were proposed to explain the synergetic behavior of Co and Mo in HDS catalysts [2-5], but it was only after the use of Mössbauer Emission Spectroscopy (MES) [6,7], when the structural Co-Mo-S model gained wide acceptance. From MES and extended X-ray absorption fine structure (EXAFS) measurements [8,9] it followed that the Co sulfide species that are located at the MoS_2 crystallite edges differ in particle size and/or ordering.

The high thiophene HDS activity observed for carbon-supported Co (Co/C) catalysts comparable with that of CoMo/C catalysts led initially to the assignment of the activity to the Co sites [10-13]. In such model studies, carbon was chosen as a support because of the absence of strong metal (sulfide)-support interactions. Subsequent MES experiments have indeed shown the presence of Co-sulfide particles with parameters similar to the Co-Mo-S doublet in Co/C catalysts after sulfidation at 373 K [14,15]. Upon further sulfidation at 673 K, the Co particles in the Co/C catalyst sintered to a Co_9S_8-type phase, whereas CoMo/C exhibited a Co-Mo-S doublet with a relatively small quadrupole splitting (Q.S.). EXAFS measurements [16] confirmed the similarities of the Co species in Co/C and CoMo/C catalysts sulfided at 373 K and suggested that the activity of CoMo/C catalysts is due to highly dispersed CoS_x particles stabilized by the secondary support MoS_2. The main function of MoS_2 is to hinder the sintering of Co sulfide species. A structural model was proposed in which part of the Co atoms four-fold coordinated by sulfur are present as very small Co_9S_8-type structures, while the rest of the Co atoms, having a sixfold sulfur coordination, are situated in a site in contact with Mo [16,17].

This model contradicted the previous observations that the activity of CoMo/C catalysts should be related to a new reaction site associated with the Co-Mo-S structures [18,19]. In support of that view are isotopic exchange experiments (H_2-D_2 equilibration) [20] that indicate the special reactivity of the sulfur atoms that bridge between Co^{2+} and Mo^{4+}. In addition, a promoting effect of Co_9S_8 has also been associated with the fact that bulk Co sulfides can act as a support for highly dispersed Co-promoted MoS_2, as observed in unsupported CoMo sulfide catalysts [21]. A square-pyramidal model, in which fivefold sulfur-coordinated Co atoms in tetragonal pyramid of sulfurs are situated on the MoS_2 edges, was also suggested for the Co-Mo-S phase [22,23].

For alumina-supported catalysts two types of Co-Mo-S structures were proposed [24]: an incompletely sulfided Type I phase with some residual linkages to the subjacent support and a highly active Type II Co-Mo-S, where all such Mo-O-Al linkages are completely sulfided. The high activity of CoMo/C as compared to γ-alumina-supported catalysts is believed to originate from a different interaction of the sulfided phase with the carbon support, leading to Type II Co-Mo-S structures [25]. The use of complexing agents also lead to preferential formation of the Type II Co-Mo-S phase by retarding sulfidation of the Co atoms until complete Mo sulfidation [26,27] and decreasing Mo-support interactions [28]. The Type I and II Co-Mo-S phases cannot be distinguished by MES parameters from room-temperature spectra [29], although some information on the support interaction strength of these structures can be obtained from measurements of the Debye temperature [13,24].

An additional positive effect of the carbon support on the activity of HDS catalysts [30] was explained in terms of its high-surface area resulting in a high dispersion of the active phase [31], controlled pore volume, reduced deactivation via coke formation and resistance to nitrogen poisoning [32-34]. The important advantage of the carbon-supported catalysts is that the oxidic precursors can be fully reduced to their active sulfide form, because of the weak metal-support interactions [35]. The major drawbacks in the utilization of carbon support in hydrotreating catalysts are the extensive microporosity – resulting in micropore blocking at high Mo contents [36,37] and, in terms of industrial relevance, the low density of carbon materials.

In this study the applicability of the Co-Mo-S structural model is evaluated for carbon-supported catalysts activated under high-pressure conditions. Using a systematic temperature stepwise investigation of the sulfidation of CoMo/C catalysts, the stability of the Co-containing phases is investigated at elevated pressures. The results are compared with the ones obtained with catalysts having the same metal loadings and prepared in the same way, sulfided at atmospheric pressure [38] and with alumina-supported catalysts sulfided at high pressure [39]. The formation and evolution of the active phase after the different applied treatments are followed using MES and EXAFS. The characterization of the catalysts is completed by catalytic DBT HDS activity measurements.

5.2 Experimental

Four CoMo catalysts were prepared by pore-volume impregnation using activated carbon (Norit RX3-extra, BET surface area 1197 $m^2 \cdot g^{-1}$, pore volume 1.0 $ml \cdot g^{-1}$, particle size 0.5-0.85 mm) as support material. Aqueous solutions of cobalt nitrate $Co(NO_3)_2 \cdot 6H_2O$ (Merck p.a.) and ammonium heptamolybdate $(NH_4)_6Mo_7O_{24} \cdot 4H_2O$ (Merck, min. 99.9%) were used in a two-step impregnation procedure, Mo being introduced first. About 50 MBq ^{57}Co as an aqueous solution of $Co(NO_3)_2 \cdot 6H_2O$ was added to the Co-containing impregnation solution, for the MES measurements. After introduction of Mo, the catalysts were dried in static air at 383 K for 16 h. Subsequently, Co was introduced and the catalysts were left in static air at 293 K for 16 h.

Four CoMo/C catalysts contained 7 wt% Mo, two of them having high promoter-loading of 2.25 wt% Co (Co(2.25)Mo(7)/C), one catalyst having 1.3 wt% Co (Co(1.3)Mo(7)/C) and one with a low promoter loading of 0.04 wt% Co (Co(0.04)Mo(7)/C). The catalyst loadings are given relative to the support material and are calculated from the impregnation solutions. The applied sulfidation treatment is denoted by the postfix (S, x MPa, y K, z h), indicating that during the experiment the catalyst is linearly heated to y K at x MPa in 1 h and kept at that temperature for z h (or 1h if z is not mentioned).

The experimental method applied for the MES measurements is presented in section 2.1.4 and the details of the EXAFS measurements and analysis are available in section 2.2.2. Gas-phase DBT HDS experiments were carried out in a high-pressure stainless steel reactor according to the method described in section 2.4.

5.3 Results

5.3.1 MES results

The Mössbauer spectra of the Co(2.25)Mo(7)/C catalyst sulfided at 4 MPa are presented in Figure 5.1 and the resulting MES parameters are shown in Table 5.1. The spectrum of the fresh catalyst and the spectrum of the catalyst exposed to the sulfidation mixture before increasing the pressure are also presented.

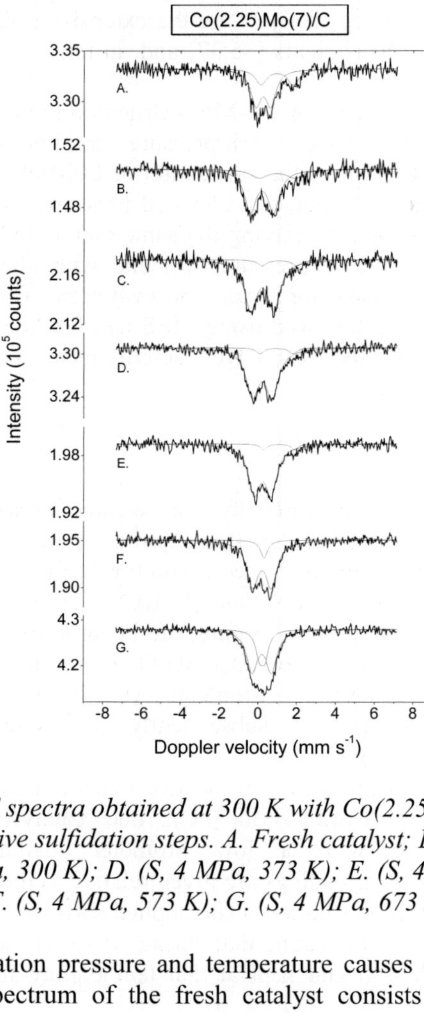

Figure 5.1 *MES spectra obtained at 300 K with Co(2.25)Mo(7)/C catalyst after various successive sulfidation steps. A. Fresh catalyst; B. (S, 0.1 MPa, 300 K); C. (S, 4 MPa, 300 K); D. (S, 4 MPa, 373 K); E. (S, 4 MPa, 473 K); F. (S, 4 MPa, 573 K); G. (S, 4 MPa, 673 K).*

Increasing the sulfidation pressure and temperature causes clear changes in the MES spectrum shape. The spectrum of the fresh catalyst consists of a quadrupole doublet indicating a low-spin 2+ or a 3+ phase and a second contribution consistent with Co^{2+} in the high-spin state [40]. Exposure of the fresh catalyst to the H_2S/H_2 gas mixture has an influence on the catalyst even at room temperature. The high-spin 2+ contribution decreases, but it does not disappear completely from the spectra until sulfidation at 573 K.

After treatment at 300 K, a second contribution with IS of 0.18 mm s^{-1} and QS of 1.21 mm s^{-1} dominates the spectrum. The QS of this doublet remains constant after sulfidation at room temperature and 4 MPa, but changes strongly upon a further increase of the sulfidation temperature. The QS value decreases to 0.89 mm s^{-1} after high-pressure sulfidation at 373 K and 473 K and then increases to 1.02 mm s^{-1} upon increasing the temperature to 673 K. Concomitant with the complete disappearance of the high-spin 2+ doublet, the spectrum of Co(2.25)Mo(7)/C shows the presence of a new doublet, which after sulfidation at 673 K has an IS of 0.23 mm s^{-1} and a QS of 0.29 mm s^{-1}. This new doublet, assigned to sintered Co-sulfide species, is observed in the spectrum already after treatment at 573 K, where, because of its relative small spectral contribution, the MES parameters could not be determined precisely, and the resulted absorption line was fitted with only one singlet (see Figure 5.1 F).

The Mössbauer spectra of the Co(0.04)Mo(7)/C catalyst sulfided at 4 MPa are presented in Figure 5.2 and the resulting MES parameters are shown in Table 5.2. The spectrum of the freshly prepared catalyst shows features similar with those of the Co(2.25)Mo(7)/C catalyst. The high-spin 2+ contribution almost doubles after exposing the catalyst to the reaction mixture, and it disappears from the spectrum upon sulfidation at 473 K.

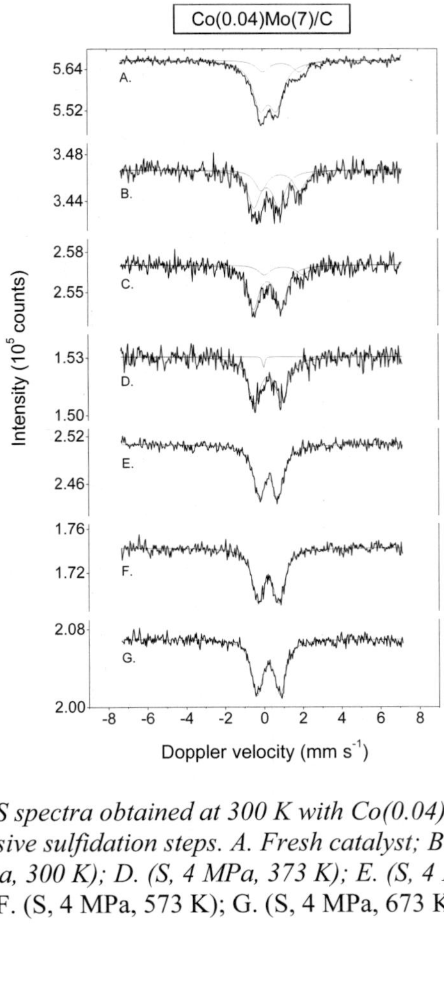

Figure 5.2 *MES spectra obtained at 300 K with Co(0.04)Mo(7)/C catalyst after various successive sulfidation steps. A. Fresh catalyst; B. (S, 0.1 MPa, 300 K); C. (S, 4 MPa, 300 K); D. (S, 4 MPa, 373 K); E. (S, 4 MPa, 473 K);* F. (S, 4 MPa, 573 K); G. (S, 4 MPa, 673 K).

***Table 5.1** MES parameters of Co(2.25)Mo(7)/C catalyst after sulfidation treatment*

T_s K	P MPa	I.S. mm s^{-1}	Q.S. mm s^{-1}	Γ mm s^{-1}	A %	I.S. mm s^{-1}	Q.S. mm s^{-1}	Γ mm s^{-1}	A %	I.S. mm s^{-1}	Q.S. mm s^{-1}	Γ mm s^{-1}	A %
		"Co – oxide"				"Co – sulfide"				"High-spin 2+"			
Fresh		0.25	0.77	0.65	66					0.96	1.67	0.72	34
300	0.1					0.18	1.21	0.74	77	0.96	1.85	1.11	23
300	4					0.20	1.20	0.78	92	1.08	1.78	0.56	8
373	4					0.21	1.03	0.80	92	0.94	1.71	0.50	8
473	4					0.24	0.89	0.74	93	1.08	1.6	0.40	7
						"Co – sulfide"				Co_9S_8			
573	4					0.20	0.98	0.72	88	0.30	-	0.46	12
673	4					0.20	1.02	0.68	70	0.23	0.29	0.47	30

***Table 5.2** MES parameters of Co(0.04)Mo(7)/C catalyst after sulfidation treatment*

T_s K	P MPa	I.S. mm s^{-1}	Q.S. mm s^{-1}	Γ mm s^{-1}	A %	I.S. mm s^{-1}	Q.S. mm s^{-1}	Γ mm s^{-1}	A %	I.S. mm s^{-1}	Q.S. mm s^{-1}	Γ mm s^{-1}	A %
		"Co – oxide"				"Co – sulfide"				"High-spin 2+"			
Fresh		0.31	0.84	0.90	80					0.97	1.88	0.81	20
300	0.1					0.20	1.30	0.76	63	0.92	1.97	0.76	37
300	4					0.23	1.41	0.70	80	1.07	1.98	0.76	20
373	4					0.25	1.40	0.76	95	1.18	2.30	0.50	5
473	4					0.26	0.96	0.67	100				
573	4					0.23	1.06	0.72	100				
673	4					0.22	1.19	0.67	100				

Again, the QS of the remaining doublet decreases to 0.96 mm s^{-1} during treatment up to 473 K, while it increases again upon further sulfidation to 673 K. It is observed that the QS value of this doublet, after sulfiding at 673 K and 4 MPa, is larger than the one obtained with the Co(2.25)Mo(7)/C catalyst treated in the same conditions. A contribution of less dispersed Co-sulfide particles as evident from the spectra of the high promoter-loading catalyst is not observed in Co(0.04)Mo(7)/C.

The MES results obtained with the Co(2.25)Mo(7)/C catalyst sulfided at atmospheric pressure and then subjected to a high-pressure treatment (Figure 5.3) are presented in Table 5.3. The high-spin 2+ contribution completely disappears from the spectrum after exposing the catalyst to the sulfidation mixture at atmospheric pressure. The Q.S. value of the remaining spectral contribution slightly increases after sulfidation at 573 K and no sintering of the Co-sulfide species takes place after treatment at 673 K and 0.1 MPa. After the high-pressure activation at 673 K, a large contribution of a small doublet with an IS of 0.25 mm s^{-1} – assigned to the Co_9S_8-like species – is observed in the spectrum.

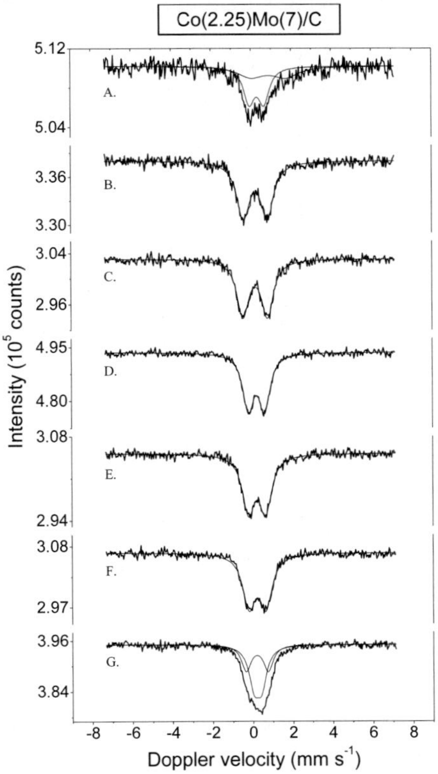

Figure 5.3 *MES spectra obtained at 300 K with Co(2.25)Mo(7)/C catalyst after various successive sulfidation steps. A. Fresh catalyst; B. (S, 0.1 MPa, 300 K); C. (S, 0.1 MPa, 373 K); D. (S, 0.1 MPa, 473 K); E. (S, 0.1 MPa, 573 K); F. (S, 0.1 MPa, 673 K); G. (S, 4 MPa, 673 K).*

The evolution of the Q.S. values for the Co(2.25)Mo(7)/C catalyst upon activation under atmospheric and high-pressure conditions is compared in Figure 5.4.

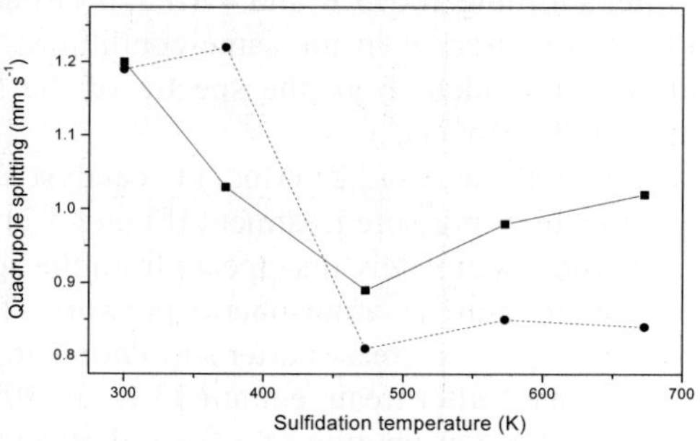

Figure 5.4 *Q.S. values of the Co-sulfide doublet obtained with Co(2.25)Mo(7)/C catalyst at 0.1 MPa (●) and at 4 MPa (■) after stepwise sulfidation at the indicated temperatures.*

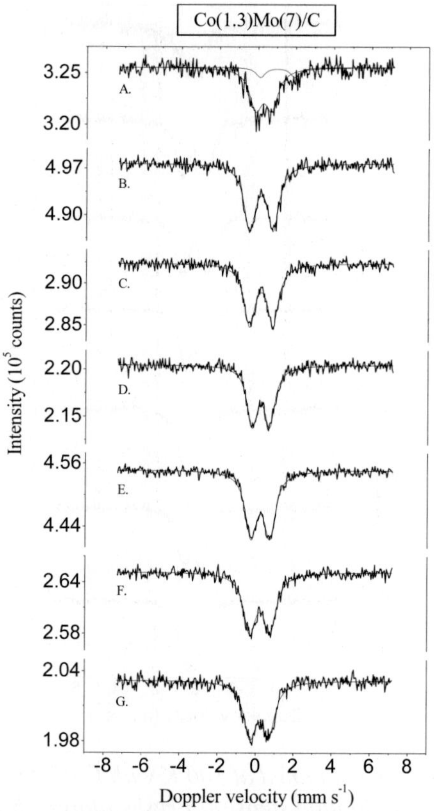

Figure 5.5 *MES spectra obtained at 300 K with Co(1.3)Mo(7)/C catalyst after various successive sulfidation steps. A. Fresh catalyst; B. (S, 0.1 MPa, 300 K); C. (S, 0.1 MPa, 373 K); D. (S, 0.1 MPa, 473 K); E. (S, 0.1 MPa, 573 K); F. (S, 0.1 MPa, 673 K); G. (S, 4 MPa, 673 K).*

***Table 5.3** MES parameters of Co(2.25)Mo(7)/C catalyst after sulfidation treatment*

T_s K	P MPa	IS mm s^{-1}	QS mm s^{-1}	Γ mm s^{-1}	A %	IS mm s^{-1}	QS mm s^{-1}	Γ mm s^{-1}	A %	IS mm s^{-1}	QS mm s^{-1}	Γ mm s^{-1}	A %
		"Co – oxide"				"Co – sulfide"				"High-spin 2+"			
Fresh		0.29	0.73	0.66	59					0.88	1.74	0.59	41
300	0.1					0.21	1.19	0.77	100				
373	0.1					0.18	1.22	0.72	100				
473	0.1					0.22	0.81	0.67	100				
573	0.1					0.23	0.85	0.73	100				
673	0.1					0.23	0.84	0.75	100				
						"Co – sulfide"				Co_9S_8			
673	4					0.21	0.98	0.66	63	0.25	0.29	0.46	37

***Table 5.4** MES parameters of Co(1.3)Mo(7)/C catalyst after sulfidation treatment*

T_s K	P MPa	IS mm·s^{-1}	QS mm s^{-1}	Γ mm s^{-1}	A %	IS mm s^{-1}	QS mm s^{-1}	Γ mm s^{-1}	A %	IS mm s^{-1}	QS mm s^{-1}	Γ mm s^{-1}	A %
		"Co – oxide"				"Co – sulfide"				"High-spin 2+"			
Fresh		0.29	0.84	0.81	84					1.02	1.82	0.59	16
300	0.1					0.19	1.24	0.73	100				
373	0.1					0.20	1.23	0.70	100				
473	0.1					0.21	0.91	0.67	100				
573	0.1					0.21	0.98	0.73	100				
673	0.1					0.21	1.05	0.79	100				
673	4					0.21	0.97	0.80	100				

Figure 5.5 shows the MES spectra obtained with the Co(1.3)Mo(7)/C catalyst sulfided at atmospheric pressure and then subjected to a high-pressure treatment. The corresponding fit parameters are presented in Table 5.4. Again, the high-spin 2+ contribution completely disappears from the spectrum after treatment at 300 K and atmospheric pressure. The Q.S. of the remaining doublet increases after sulfidation at 573 K. The resulting doublet has an I.S. value of 0.21 mm s^{-1} and a Q.S. value of 1.05 mm s^{-1}, parameters that are characteristic for Co-Mo-S. After activation at 673 K and 4 MPa, the Q.S. decreases slightly, but no segregation of a Co-sulfide species takes place.

5.3.2 Mo EXAFS

The EXAFS fit parameters at the Mo K edge of Co(2.25)Mo(7)/C catalyst sulfided at intermediate temperatures, under atmospheric and high-pressure conditions, are collected in Table 5.5.

***Table 5.5** Mo K-edge EXAFS parameters of Co(2.25)Mo(7)/C catalyst after different sulfidation treatments*

Treatment	N -	R Å	$\Delta\sigma^2$ $Å^2$ $\cdot 10^{-3}$	ΔE_0 eV	N -	R Å	$\Delta\sigma^2$ $Å^2$ $\cdot 10^{-3}$	ΔE_0 eV	N -	R Å	$\Delta\sigma^2$ $Å^2$ $\cdot 10^{-3}$	ΔE_0 eV
	Mo-S				Mo-Mo				Mo-Co			
S, 4 MPa, 473 K	5.7	2.41	6.3	2.3	2.0	2.80	9.7	-6.9				
S, 4 MPa, 573 K	5.4	2.42	3.0	-0.3	2.4	3.15	5.6	3.7	1.0	2.77	9.9	9.5
S, 4 MPa, 673 K	5.8	2.42	1.7	0.6	3.9	3.16	4.2	1.6	0.6	2.75	6.5	9.8
S, 0.1 MPa, 473 K	5.1	2.41	9.8	2.6	0.5	2.78	4.3	-3.1				
S, 0.1 MPa, 573 K	4.9	2.41	3.5	0.4	2.2	3.14	5.5	4.4				
S, 0.1 MPa, 673 K	5.3	2.42	3.2	0.6	2.4	3.15	3.9	1.6				
S, 0.1 MPa + S, 4 MPa, 673 K	5.7	2.42	1.6	2.2	4.0	3.15	4.0	3.7	0.5	2.75	7.2	8.7

The spectrum of the sample sulfided at 473 K and 4 MPa was fitted with a Mo-S contribution at 2.41 Å and a Mo-Mo contribution at 2.8 Å. The Mo-Mo distance is smaller than the typical distance of 3.16 Å in MoS_2 [41]. A high Mo-S coordination number is observed at this temperature, indicating a possible overlap of Mo-O and Mo-S shells or a different Mo valence [42,43]. After increasing the sulfidation temperature to 573 K and 673 K, the Mo-S and Mo-Mo interatomic distances are in agreement with the structural parameters of bulk MoS_2. A Mo-Co contribution is observed at a distance of about 2.77 Å after increasing the temperature to 573 K, with $N_{Mo\text{-}Co}$ slightly decreasing after treatment at 673 K and 4 MPa.

In comparison, the fit parameters for the sample sulfided at 0.1 MPa indicate a slower sulfidation as follows from the lower Mo-S and Mo-Mo coordination numbers at the intermediate temperatures. In the final state at 673 K, the low Mo-Mo and Mo-S coordination numbers suggest that the particle size is smaller and/or more disordered. In line with this, no Mo-Co coordination is observed. Such a contribution appears upon increasing the sulfidation pressure to 4 MPa. The structural parameters of this sample are very similar to those of Co(2.25)Mo(7)/C directly sulfided at 4 MPa.

The fit results of Mo K-edge EXAFS measurements of Co(1.3)Mo(7)/C catalyst sulfided at atmospheric pressure and then subjected to a high-pressure treatment, are collected in Table 5.6. After sulfidation at 473 K and 0.1 MPa, the Mo-S distance of 2.46 Å is larger than the corresponding distance of 2.41 Å in crystalline MoS_2, whereas the Mo-Mo distance is again smaller than the typical distance of 3.16 Å in MoS_2. After increasing the temperature, the presence of MoS_2-like species is evident at all the intermediate steps. Again, the Mo-Co contribution is observed only after the subsequent treatment at 4 MPa.

Table 5.6 *Mo K-edge EXAFS parameters of Co(1.3)Mo(7)/C catalyst after different sulfidation treatments*

Treatment	N -	R Å	$\Delta\sigma^2$ Å^2 $\cdot 10^{-3}$	ΔE_0 eV	N -	R Å	$\Delta\sigma^2$ Å^2 $\cdot 10^{-3}$	ΔE_0 eV	N -	R Å	$\Delta\sigma^2$ Å^2 $\cdot 10^{-3}$	ΔE_0 eV
	Mo-S				Mo-Mo				Mo-Co			
S, 0.1 MPa, 473 K	5.7	2.46	5.4	0.8	1.3	2.73	4.5	4.6				
S, 0.1 MPa, 573 K	4.6	2.41	3.7	1.5	2.3	3.13	6.2	4.1				
S, 0.1 MPa, 673 K	5.3	2.41	2.7	2.6	2.9	3.15	4.1	2.6				
S, 0.1 MPa + S, 4 MPa, 673 K	5.7	2.42	2.1	2.5	3.6	3.15	4.1	4.6	0.5	2.75	8.6	9.8

5.3.3 Co EXAFS

The Co K-edge EXAFS fit parameters of the sulfided Co(2.25)Mo(7)/C catalyst are listed in Table 5.7. After treatment at 4 MPa, the spectra were fitted with a Co-S contribution at a distance of 2.22 Å and two Co-Co contributions at 2.6 Å and at 3.3 Å, distances that are different from those corresponding to bulk Co_9S_8 (2.51 Å and 3.51 Å, respectively). A Co-Mo contribution at 2.8 Å could be included in the spectra measured after treatment at 573 K and 673 K.

Upon activation of the same catalyst at atmospheric pressure, the Co-S coordination number is higher for the sample sulfided at 673 K compared to that of the catalyst sulfided at the same temperature and high-pressure. After the subsequent treatment at 673 K and 4 MPa, $N_{Co\text{-}S}$ decreases from 5.8 to 5.2 and the Co-Mo contribution is observed in the spectrum.

The fit results of Co K-edge EXAFS measurements of Co(1.3)Mo(7)/C catalyst sulfided at atmospheric pressure and then subjected to a high-pressure treatment, are collected in Table 5.8. The Co-S coordination number increases from 5.3 to 5.6 upon increasing the sulfidation pressure from 0.1 to 4 MPa at 673 K. Again, the Co-Mo contribution is observed only after the subsequent treatment at 4 MPa.

Table 5.7 *Co K-edge EXAFS parameters of Co(2.25)Mo(7)/C catalyst after different sulfidation treatments*

Treatment	N -	R Å	$\Delta\sigma^2$ Å^2 $\cdot 10^{-3}$	ΔE_0 eV	N -	R Å	$\Delta\sigma^2$ Å^2 $\cdot 10^{-3}$	ΔE_0 eV	N -	R Å	$\Delta\sigma^2$ Å^2 $\cdot 10^{-3}$	ΔE_0 eV	N -	R Å	$\Delta\sigma^2$ Å^2 $\cdot 10^{-3}$	ΔE_0 eV
	Co-S				Co-Co (1)				Co-Mo				Co-Co (2)			
S, 4 MPa, 473 K	4.8	2.22	3.4	3.0	1.3	2.52	4.0	-7.9					1.0	3.43	6.4	-9.7
S, 4 MPa, 573 K	5.6	2.23	2.1	-4.5	1.1	2.60	9.5	0.8	1.1	2.80	7.7	0.8	0.7	3.30	9.5	7.9
S, 4 MPa, 673 K	5.3	2.23	1.3	-5.8	1.3	2.60	3.5	-2.5	0.6	2.80	9.9	-9.6	1.2	3.30	9.8	3.0
S, 0.1 MPa, 473 K	5.1	2.22	1.4	2.3	2.7	2.56	6.0	8.5					2.2	3.31	9.4	-5.0
S, 0.1 MPa, 573 K	5.5	2.22	2.9	-1.0	1.4	2.57	6.6	-9.5					0.9	3.24	9.9	8.4
S, 0.1 MPa, 673 K	5.8	2.23	2.7	-4.3	0.7	2.60	0.1	-1.5					1.3	3.27	9.0	8.2
S, 0.1 MPa + S, 4 MPa, 673 K	5.2	2.23	1.9	-4.5	1.5	2.60	9.8	0.8	0.7	2.80	9.5	-8.0	1.4	3.26	8.1	9.4

Table 5.8 *Co K-edge EXAFS parameters of Co(1.3)Mo(7)/C catalyst after different sulfidation treatments*

Treatment	N -	R Å	$\Delta\sigma^2$ Å^2 $\cdot 10^{-3}$	ΔE_0 eV	N -	R Å	$\Delta\sigma^2$ Å^2 $\cdot 10^{-3}$	ΔE_0 eV	N -	R Å	$\Delta\sigma^2$ Å^2 $\cdot 10^{-3}$	ΔE_0 eV	N -	R Å	$\Delta\sigma^2$ Å^2 $\cdot 10^{-3}$	ΔE_0 eV
	Co-S				Co-Co (1)				Co-Mo				Co-Co (2)			
S, 0.1 MPa, 473 K	5.1	2.22	3.6	0.4	2.0	2.56	9.5	-3.4					1.2	3.29	9.9	9.5
S, 0.1 MPa, 573 K	5.2	2.23	3.4	-0.8	1.3	2.57	5.8	-3.0					1.3	3.29	9.6	9.5
S, 0.1 MPa, 673 K	5.3	2.23	3.6	-1.1	1.2	2.57	5.6	-2.2					1.2	3.29	9.7	9.5
S, 0.1 MPa + S, 4 MPa, 673 K	5.6	2.23	4.2	-1.2	1.4	2.60	9.5	-1.4	1.1	2.80	9.6	-9.7	0.9	3.35	9.8	9.8

5.3.4 Dibenzothiophene HDS

The DBT HDS activities of the Co(2.25)Mo(7)/C catalyst activated at 673 K and different pressures are shown in Table 5.9. The rate constant (k_{HDS}), measured after 8 hours at 3 MPa and 573 K, is expressed per kg of Mo. The apparent activation energies (E_{act}) and the pre-exponential factors (ν_{pre}) were determined in the 533-573 K temperature interval.

***Table 5.9** DBT HDS activities, apparent activation energies and pre-exponential factors of Co(2.25)Mo(7)/C catalyst after different sulfidation treatments*

Treatment	k_{HDS} *(mol DBT·Kg $_{Mo}^{-1}$·s^{-1}) ·10^{-2}*	E_{act} *(kJ·mol^{-1})*	ν_{pre} *(mol·kg^{-1}·s^{-1})*
S, 4 MPa, 673 K	10.1	120	$9\cdot10^{9}$
S, 0.1 MPa, 673 K	3.0	81	$6.7\cdot10^{5}$
S, 0.1 MPa, 673 K + S, 4 MPa, 673 K	8.1	109	$7\cdot10^{8}$

It is found that the sulfidation pressure has a large influence on the DBT HDS activity; the reaction rate constant is considerably higher after sulfidation at 4 MPa compared to atmospheric pressure activation. The higher activity for the high-pressure sulfided samples goes with a considerably higher apparent activation energies and correspondingly higher pre-exponential factors. The catalytic results for Co(1.3)Mo(7)/C sample are summarized in Table 5.10. A higher activity was observed with the sample sulfided at 673 K and 0.1 MPa compared to that of the catalyst having a high Co loading, which was activated in the same conditions. The reaction rate constant increases for the sample subsequently treated at 4 MPa, but to a lesser extent than in the Co(2.25)Mo(7)/C case.

***Table 5.10** DBT HDS activities, apparent activation energies and pre-exponential factors of Co(1.3)Mo(7)/C catalyst after different sulfidation treatments*

Treatment	k_{HDS} *(mol DBT·Kg $_{Mo}^{-1}$·s^{-1}) ·10^{-2}*	E_{act} *(kJ·mol^{-1})*	ν_{pre} *(mol·kg^{-1}·s^{-1})*
S, 0.1 MPa, 673 K	3.8	111	$4.6\cdot10^{8}$
S, 0.1 MPa, 673 K + S, 4 MPa, 673 K	6.0	127	$2.3\cdot10^{10}$

5.4 Discussion

The first *in situ* high-pressure MES studies on CoMo/Al_2O_3 catalysts have shown that under sulfidation conditions more close to industrial practice the active phase can still be well described in terms of Co-Mo-S structures [39,44]. A pivotal question is whether in high-pressure sulfided carbon-supported CoMo catalysts Co is present in a unique phase as indicated in the Co-Mo-S model, or whether the highly dispersed Co-sulfide particles (CoS_x) stabilized by MoS_2 are formed.

The spectra of all fresh samples consist of two quadrupole doublets. The main spectral contribution, with an I.S. of 0.25 - 0.31 mm s^{-1}, is assigned to Co-oxide species. The second doublet, representing a high-spin 2+ phase, must be assigned to oxygen containing species and to a Co-sulfate type phase at the intermediate sulfidation steps, as suggested by previous EXAFS measurements [16]. The complete sulfidation of this intermediate high-spin 2+ contribution takes a longer time when the samples are treated directly at 4 MPa (Figures 5.1 and 5.2), pointing to a stronger Co-Mo interaction under high-pressure conditions.

For the catalyst with a lower Co/Mo ratio, exposure to the H_2S/H_2 gas mixture at room temperature leads to an increase in the spectral contribution of the high-spin 2+ phase. This can be explained by removal of water associated with the high-spin 2+ species in the fresh catalyst, causing an increased interaction of these structures with the support [48]. The larger amount of unsulfided high-spin 2+ phase present at this stage compared with the high promoter-loading catalyst can also be understood in terms of formation of the intermediate Co-sulfate type phase, that is more difficult to sulfide in low promoter-loading catalysts under atmospheric pressure conditions [38]. The MES parameters obtained after sulfidation at room temperature are different from those of their fresh counterparts. The main spectral contribution is assigned to Co-sulfide species as indicated by Crajé et al [49]. The preferential adsorption of Co atoms on Mo during the preparation process allows the Co sulfidation to start already at room temperature, even for the low promoter-loading catalyst [38].

During treatment at 573 K and 673 K the Q.S. of the dominant doublet increases, indicating a decrease in particle size and/or ordering of the Co-sulfide species [16]. The MoS_2 crystallites are formed between 523 K and 573 K and their sizes are found to increase with increasing sulfiding temperature to 573 K and 673 K [41,42]. Thus, the observed increase in the Q.S. after treatment at 573 K is understood in terms of redispersion of the Co-sulfide particles over the edges of newly formed MoS_2 crystallites. A less pronounced redispersion of the Co-sulfide species at 573 K is noted for the Co(2.25)Mo(7)/C and Co(1.3)Mo(7)/C catalysts sulfided at 0.1 MPa (Figures 5.4 and 5.5), confirming the weaker Co-Mo interaction under these conditions.

As observed in Figure 5.1, the sulfidation of Co(2.25)Mo(7)/C catalyst can be well described in terms of formation of a dominant active species with parameters of the Co-Mo-S phase. Upon sulfidation at 4 MPa and 673 K, the active phase exhibits MES parameters (I.S. = 0.20 mm s^{-1}; Q.S. = 1.02 mm s^{-1}), which fall within the range of parameters that have been suggested for the Co-Mo-S structure (I.S. = 0.22 mm s^{-1}; Q.S. = 1.0 - 1.3 mm s^{-1}) [25]. Concomitant with Co-Mo-S formation, separate Co_9S_8-like structures are formed as previously observed with a similarly prepared uncalcined alumina-supported catalyst at the same metal loadings (Chapter 4). In that case, however, the weak Co-Mo interaction led to the partial formation of small CoS_x particles having MES parameters that resemble the Co-Mo-S parameters [38]. Complete sintering is usually hindered by MoS_2, but upon more severe sulfidation treatments, the CoS_x structures grow and the resulting Co-sulfide species exhibit structural parameters that approach those of crystalline Co_9S_8 [16]. These findings confirm the observations that the differences between the Co-Mo-S model and previous catalyst models can be related to the fact that the promoter atoms are not always all present in the Co-Mo-S structures [45,46]. Depending on the choice of activation conditions, highly dispersed Co_9S_8-type structures can also be formed.

Already after sulfidation at 573 K and 4 MPa the contribution assigned to sintered Co-sulfide species is observed in the spectrum of the high-loading catalyst (Figure 5.1, F). Upon treatment at 673 K the resulting doublet has a Q.S. value of 0.29 mm s^{-1}, very close to the value for crystalline Co_9S_8 (Q.S. = 0.26 mm s^{-1}) [6]. Co_9S_8 formation takes place faster than in the calcined and uncalcined alumina-supported catalysts [Chapters 3,4] - because of the lower metal-support interaction.

The Co_9S_8-like doublet is not observed in the spectra of Co(2.25)Mo(7)/C catalyst activated under atmospheric pressure conditions (Figure 5.3) which is attributed to a slower rate of Co sulfidation under these conditions. The relatively low Q.S. value of the Co-sulfide doublet obtained after treatment at 673 K and 0.1 MPa (Q.S. = 0.84 mm s^{-1}) points to the presence of a considerable fraction of small CoS_x particles. These CoS_x species sinter to larger particles (Co_9S_8-like) after subsequent high-pressure sulfidation at 673 K, leaving about 63 % of the Co atoms in the active Co-Mo-S structures (Figure 5.3, G). The Co_9S_8 doublet is not observed in the spectra of sulfided Co(0.04)Mo(7)/C and Co(1.3)Mo(7)/C samples, since the Co_9S_8 formation is shifted to higher temperatures with decreasing Co content [47].

The Co K-edge EXAFS results obtained with the Co(2.25)Mo(7)/C catalyst (Table 5.7) show that the Co atoms are sulfided already after exposure to the H_2S/H_2 gas mixture at room temperature under atmospheric and high-pressure conditions. After treatment at 473 K and 4 MPa, the resulted Co-S (2.22 Å) and Co-Co(1) (2.52 Å) coordination distances are closer to the values previously reported for Co_9S_8 [17, 22]. Also, the longer Co-Co(2) coordination distance (3.43 Å) – closer to the corresponding distance in Co_9S_8 – indicates the formation of larger Co-sulfide species as a result of the higher rate of Co sulfidation at 4 MPa.

High Co-S coordination numbers (5.5 – 5.8) are measured after sulfidation at 573 K under atmospheric and high-pressure conditions and after activation at 673 K and 0.1 MPa, confirming that no large amounts of Co_9S_8 are formed after these treatments. The MES results clearly showed large contributions from the Co_9S_8-like doublets only after activation under high-pressure conditions at 673 K (Figure 5.1, G and Figure 5.3, G). Very disordered Co-sulfide species, with coordination distances and numbers different from those of crystalline cobalt sulfide systems, are formed at the intermediate temperatures, as previously observed with CoMo/C [16] and with $NiMo/SiO_2$ [26] catalysts. These CoS_x structures are formed as soon as the catalysts are exposed to the sulfidation mixture and after activation at 573 and 673 K only part of the Co atoms will participate in the Co-Mo-S species. The partial redispersion of the Co-sulfide particles at the temperature of MoS_2 crystallites formation, is confirmed by the decrease of the Co-Co(1) and Co-Co(2) coordination numbers at 573 K. A similar redispersion model has been proposed for NiS_x particles in NiW catalysts [50].

The EXAFS data analysis of the sample sulfided at 573 K and 4 MPa reveals the presence of a Co-Mo contribution at 2.8 Å, in good agreement with the previously reported values [16,51]. This is in line with the conclusion that Co-Mo-S structures are formed at this temperature. For the sample treated first under atmospheric pressure conditions, a small Co-Mo coordination shell is observed only after sulfidation at 673 K and 4 MPa. The higher Co-Mo coordination numbers obtained with the catalysts activated at 4 MPa point in the direction of Type II Co-Mo-S phase formation upon increasing sulfidation pressure. It has generally been assumed that, because of the inertness of the carbon support, Type II Co-Mo-S is formed even at 0.1 MPa and normal sulfidation temperature, but this seems to up for discussion now (*vide infra*). The Co species become better defined upon sulfidation at higher pressure. Because of the partial formation of CoS_x species, the observed Co-Mo coordination numbers are lower compared to those obtained with the similar calcined sample supported on alumina (Chapter 3). The decrease of the Co-S coordination numbers after the high-pressure treatments at 673 K confirms the sintering of the CoS_x structures to Co_9S_8, where Co atoms are fourfold coordinated with S.

Also the Co(1.3)Mo(7)/C catalyst sulfided first at atmospheric pressure (Table 5.8), shows the presence of a Co-Mo contribution only after treatment at 673 K and 4 MPa. This is consistent with the observation that Type I-like Co-Mo-S species are formed under atmospheric pressure conditions, as a result of an incomplete sulfidation or a Mo-carbon interaction (e.g. via oxygen surface groups). No decrease in the Co-S coordination number is observed after increasing the activation pressure at 673 K, since no Co_9S_8-type structures are formed in this sample.

The Mo EXAFS fit results for Co(2.25)Mo(7)/C sulfided under both atmospheric and high pressure conditions (Table 5.5) reveal the presence of relatively large Mo-S coordination numbers ($N_{Mo\text{-}S}$ = 5.1 – 5.7), in combination with short Mo-Mo distances (2.78 – 2.80 Å), after treatment at 473 K. Previous EXAFS studies attributed these contributions to a MoS_3-like intermediate species [43,52,53]. After increasing the sulfidation temperature to 573 K and 673 K, the Mo-S and Mo-Mo interatomic distances are in agreement with the structural parameters of MoS_2 crystallites. A lower Mo-S coordination number ($N_{Mo\text{-}S}$ = 5.3) is obtained after treatment at 673 K and 0.1 MPa, indicating that the MoS_2 edges are not completely saturated with sulfur. Previous XPS and EXAFS studies have shown for carbon-supported catalysts Mo-S coordination numbers, structural order and stacking degree similar to Type I Co-Mo-S catalysts supported on alumina [54]. A relatively strong MoS_2-carbon support interaction or physical trapping of the MoS_2 particles in the micropores of the support were proposed. Although complete sulfidation of the active species is expected to be achieved by using an inert support material, the slower rate of Mo sulfidation under atmospheric pressure conditions leads to the formation of a Type I-like Co-Mo-S phase.

The Mo-Co contributions are observed starting with the treatment at 573 K for the sample sulfided directly at 4 MPa, and only after sulfidation at 673 K and 4 MPa for the catalyst treated first under atmospheric pressure conditions – in good agreement with the Co EXAFS fit results. The higher Mo-S and Mo-Mo coordination numbers measured after sulfidation of both samples at 673 K and 4 MPa, point towards an accelerated Mo sulfidation under high-pressure conditions. The better saturation by sulfur of the MoS_2 edges and the formation of more well-ordered crystallites upon sulfidation at 4 MPa indicate the presence of the Type II Co-Mo-S species under these conditions. The fit results of Mo K-edge EXAFS measurements of Co(1.3)Mo(7)/C catalyst sulfided at atmospheric pressure and then subjected to a high-pressure treatment (Table 5.6) show sulfidation behavior similar to the high promoter-loading sample treated under the same conditions. Again, the Type II Co-Mo-S phase formation is observed upon the subsequent activation at 673 K and 4 MPa.

The DBT HDS measurements of the Co(2.25)Mo(7)/C catalyst sulfided at 673 K and different pressures (Table 5.9) show higher reaction rate constants compared to those obtained with an alumina-supported calcined sample (Chapter 3). The large activity difference has also been reported previously [55, 56] and originates from a more favorable interaction of the Co-Mo-S species with the carbon support [34]. It has been mentioned that the specific activity of the Type II Co-Mo-S structures is higher when supported on carbon than on alumina [30]. The reaction rate constants are considerably higher for the samples sulfided at 4 MPa compared to the sample that was activated only at atmospheric pressure. The ratio of the DBT HDS activities is higher than the relative activities of the Type I and II Co-Mo-S species observed in gas phase thiophene HDS, where Type II phase was twice as active than Co-Mo-S I [52,57]. Formation of CoS_x structures that block the active sites can explain the lower activity of the Type I-like species formed after sulfidation at 0.1 MPa. The activity of the sample sulfided first under atmospheric pressure and subsequently at 4 MPa is somewhat smaller compared to that of the catalyst activated directly under high-pressure conditions, because of the lower amount of Co-Mo-S active species present in the former sample as shown by the MES results.

To obtain a better understanding of the nature of Type I-like phase formed in the carbon-supported catalysts after activation at 0.1 MPa, catalytic activities in thiophene HDS were determined for the Co(2.25)Mo(7)/C sample. The experimental details of the atmospheric pressure gas-phase thiophene HDS measurements carried out are presented in section 7.2. After sulfidation at 0.1 MPa and 673 K, Co(2.25)Mo(7)/C was about twice as active in thiophene HDS as the same catalyst treated at 4 MPa (86 $mol_{thiophene} \cdot mol_{Mo}^{-1} \cdot h^{-1}$ *vs.* 39 $mol_{thiophene} \cdot mol_{Mo}^{-1} \cdot h^{-1}$). This difference is remarkable given the reverse trend in DBT HDS activity: the activity of the high-pressure sulfided Co(2.25)Mo(7)/C catalyst is more than three times larger than that of the 0.1 MPa sulfided sample. The most straightforward model that emerges is that the atmospheric pressure activation renders the presence of the active species (highly dispersed Type II Co-Mo-S) constrained in the micropores of the carbon support. The lower than expected DBT HDS activity for this atmospheric pressure sulfided sample may be explained in terms of DBT molecules that are not able to reach all the active sites. Upon sulfidation at 4 MPa and 673 K, considerable redistribution of the MoS_2 phase takes place leading to a sintered, better sulfided type II phase which is now not occluded anymore in the smallest micropores. The loss in dispersion of the MoS_2 phase and some loss of the 'Co-Mo-S' phase itself impedes the thiophene HDS activity whereas the higher accessibility of the active sites for DBT leads to a higher specific DBT HDS activity.

The DBT HDS activity obtained with the Co(1.3)Mo(7)/C catalyst sulfided at atmospheric pressure (Table 5.10) is higher compared to that of the sample having a high Co loading, because no CoS_x species are formed in the former case. The reaction rate constant increases significantly for the Co(1.3)Mo(7)/C sample subsequently treated at 4 MPa, confirming the Type II phase formation upon increasing the sulfidation pressure.

The apparent activation energies are higher for the catalysts activated also under high-pressure conditions, behavior understood in terms of increasing surface coverage by dibenzothiophene or H_2S [58]. A strong interaction of the active species with the adsorbed complex can also explain the high pre-exponential factors obtained with these samples. The large changes in the pre-exponential factors could also be related to differences in the number of active sites formed [59].

5.5 Conclusions

The sulfidation of CoMo/C catalysts under conditions more relevant to industrial practice was studied by MES, EXAFS and DBT HDS. The weak interaction of the promoter atoms with the support and also the weak Co-Mo interaction leads to a significant formation of small inactive CoS_x particles (highly dispersed Co_9S_8 type structures) in the high promoter-loading catalyst, having MES parameters that resemble the Co-Mo-S parameters. However, MES can still distinguish between the different Co-sulfides present, since smaller Q.S. values are clearly indicating the presence of CoS_x structures involving variable number of Co atoms. MoS_2 hinders the complete sintering of the CoS_x structures under atmospheric pressure conditions, but upon more severe sulfidation treatments at 4 MPa these CoS_x species sinter to Co_9S_8-type structures.

The stepwise sulfidation approach points to somewhat higher sulfidation rates of Co and Mo during high-pressure activation. This leads to the formation of more aggregated Co-sulfide species at intermediate temperatures, followed by redispersion of these species at 573 K over the edges of the newly formed MoS_2 crystallites.

The high-pressure sulfided samples have much higher DBT HDS activities than the catalysts activated at 0.1 MPa, because of the formation of Type II Co-Mo-S structures at elevated pressures. This conclusion is supported by the higher Co-Mo coordination numbers and the accelerated Mo sulfidation under high-pressure conditions, as follows from structural analysis by EXAFS. The lower rate of Mo sulfidation under atmospheric pressure conditions leads to the formation of an active phase with a Type I-character, which is defined in terms of highly dispersed Type II Co-Mo-S species constrained in the micropores of the carbon support.

References

1. Directive 2003/17/EC of the European Parliament and of the Council - amending Directive 98/70/EC relating to the quality of petrol and diesel fuels.
2. B.C. Gates, J.R. Katzer, G.C.A. Schuit, in Chemistry of Catalytic Processes, McGraw-Hill, New York, (1979) 411.
3. R.J.H. Voorhoeve, J.C.M. Stuiver, J. Catal. 23 (1971) 243.
4. A.L. Farragher, P. Cossee, in Proc. 5th Int. Congress on Catalysis, Palm Beach, ed. J.W. Hightower (North Holland, Amsterdam, 1973) 1301.
5. B. Delmon, in Proc. 3rd Int. Conference on the Chemistry and Uses of Molybdenum, Ann Arbor, ed. H.F. Barry and P.C.H. Mitchell (Climax Molybdenum Co., 1979) 73.
6. H. Topsøe, B.S. Clausen, R. Candia, C. Wivel and S. Mørup, J. Catal. 68 (1981) 433.
7. C. Wivel, R. Candia, B.S. Clausen, S. Mørup and H. Topsøe, J. Catal. 68 (1981) 453.
8. M.W.J. Crajé, V.H.J. de Beer, A.M. van der Kraan, Hyp. Int. 69 (1991) 795.
9. M.W.J. Crajé, V.H.J. de Beer, J.A.R. van Veen, A.M. van der Kraan, Appl. Catal. A 100 (1993) 97.

10. J.P.R. Vissers, V.H.J. de Beer, R. Prins, J. Chem. Soc., Faraday Trans.1 83 (1987) 2145.
11. V.H.J. de Beer, J.C. Duchet, R. Prins, J. Catal. 72 (1981) 369.
12. J.C. Duchet, E.M. van Oers, V.H.J. de Beer, R. Prins, J. Catal. 80 (1983) 386.
13. M.W.J. Crajé, V.H.J. de Beer, A.M. van der Kraan, Appl. Catal. 70 (1991) L7.
14. A.M. van der Kraan, M.W.J. Crajé, E. Gerkema, W.L.T.M. Ramselaar, V.H.J. de Beer, Appl. Catal. 39 (1988) L7.
15. A.M. van der Kraan, M.W.J. Crajé, E. Gerkema, W.L.T.M. Ramselaar, V.H.J. de Beer, Hyp. Int. 46 (1989) 567.
16. M.W.J. Crajé, S.P.A. Louwers, V.H.J. de Beer, R. Prins, A.M. van der Kraan, J. Phys. Chem. 96 (1992) 5445.
17. S.M.A.M. Bouwens, J.A.R. van Veen, D.C. Koningsberger, V.H.J. de Beer, R. Prins, J. Phys. Chem. 95 (1991) 123.
18. H. Topsøe, N.-Y. Topsøe, B.S. Clausen, 12th Iberoam. Symp. Catal., Rio de Janeiro, 2 (1990) 762.
19. M. Breysse, B.A. Bennett, D. Chadwick, M. Vrinat, Bull. Soc. Chim. Belg. 90 (1981) 1271.
20. E.J.M. Hensen, G.M.H.J. Lardinois, V.H.J. de Beer, J.A.R. van Veen, R.A. van Santen, J. Catal. 187 (1999) 95.
21. K. Inamura, R. Prins, Stud. Surf. Sci. Catal. 92 (1995) 401.
22. W. Niemann, B.S. Clausen, H. Topsøe, Catal. Lett. 9 (1990) 355.
23. S.P.A. Louwers, R. Prins, J. Catal. 133 (1992) 94.
24. R. Candia, O. Sorensen, J. Villadsen, N.-Y. Topsøe, B.S. Clausen, H. Topsøe, Bull. Soc. Chim. Belg. 93 (1984) 763.
25. H. Topsøe, R. Candia, N.-Y. Topsøe, B.S. Clausen, Bull. Soc. Chim. Belg. 93 (1984) 783.
26. L. Medici, R. Prins, J. Catal. 163 (1996) 38.
27. L. Coulier, V.H.J. de Beer, J.A.R. van Veen, J.W. Niemantsverdriet, J. Catal. 197 (2001) 26.
28. E.J.M. Hensen, V.H.J. de Beer, J.A.R. van Veen, R.A. van Santen, Catal. Lett. 84 (2002) 59.
29. J.A.R. van Veen, E. Gerkema, A.M. van der Kraan, P.A.J.M. Hendriks, H. Beens, J. Catal. 133 (1992) 112.
30. J.A.R. van Veen, E. Gerkema, A.M. van der Kraan, A. Knoester, J. Chem. Soc. Chem. Commun. (1987) 1684.
31. E.J.M. Hensen, P.J. Kooyman, Y. van der Meer, A.M. van der Kraan, V.H.J. de Beer, J.A.R. van Veen, R.A. van Santen, J. Catal. 1999 (2001) 224.
32. F. Luck, Bull. Soc. Chim. Belg. 100 (1991) 781.
33. M. Breysse, J.L. Portefaix, M. Vrinat, Catal. Today 10 (1991) 489.
34. H. Farag, D. Whitehurst, I. Mochida, Ind. Eng. Chem. Res. 37 (1998) 3533.
35. J.P.R. Vissers, C.K. Groot, E.M. van Oers, V.H.J. de Beer, R. Prins, Bull. Soc. Chim. Belg. 93 (1984) 813.
36. M. Ferrari, B. Delmon, P. Grange, Carbon 40 (2002) 497.
37. M. Ferrari, B. Delmon, P. Grange, Micropor. Mesopor. Mat. 56 (2002) 279.
38. M.W.J. Crajé, Ph.D. Thesis, Delft University of Technology, Delft, 1992, ISBN 90-73861-08-X.
39. A.I. Dugulan, M.W.J. Crajé, G.J. Kearley, J. Catal. 222 (2004) 281.

40. J.G. Steven, L. Zhe, H. Pollak, V.E. Stevens, R.H. White and J.L. Gibson, Mössbauer Handbook Minerals, Mössbauer Effect Data Center, University of North Carolina, Asheville, 1983.
41. T.G. Parham, R.P. Merrill, J. Catal. 85 (1984) 295.
42. M. de Boer, A.J. van Dillen, D.C. Koningsberger, J.W. Geus, Jpn. J. Appl. Phys. 32 (1992) 460.
43. R. Cattaneo, T. Weber, T. Shido, R. Prins, J. Catal. 191 (2000) 225.
44. A.I. Dugulan, E.J.M. Hensen, J.A.R. van Veen, Catal. Today 130 (2008) 126.
45. S. Eijsbouts, Appl. Catal. A 158 (1997) 53.
46. H. Topsøe, J. Catal. 216 (2003) 155.
47. R. Candia, B.S. Clausen, H. Topsøe, Proc. 9th Iberoam. Symp. Catal., 16-21 July 1984, Lisbon, Portugal, 1984, p. 211.
48. M.W.J. Crajé, E. Gerkema, V.H.J. de Beer, A.M. van der Kraan, Hyp. Int. 57 (1990) 1795.
49. M.W.J. Crajé, V.H.J. de Beer, A.M. van der Kraan, Bull. Soc. Chim. Belg. 100 (1991) 953.
50. E.J.M. Hensen, Y. van der Meer, J.A.R. van Veen and J.W. Niemantsverdriet, Appl. Catal. A, 322 (2007) 16.
51. S.M.A.M. Bouwens, D.C. Koningsberger, V.H.J. de Beer, S.P.A. Louwers, R. Prins, Catal. Lett. 5 (1990) 273.
52. S.P. Cramer, K.S. Liang, A.J. Jacobson, C.H. Chang, R.R. Chianelli, Inorg. Chem. 23 (1984) 215.
53. R. Cattaneo, F. Rota, R. Prins, J. Catal. 199 (2001) 318.
54. S.M.A.M. Bouwens, F.B.M. van Zon, M.P. van Dijk, A.M. van der Kraan, V.H.J. de Beer, J.A.R. van Veen, D.C. Koningsberger, J. Catal. 146 (1994) 375.
55. E.J.M. Hensen, V.H.J. de Beer, J.A.R. van Veen and R.A. van Santen, J. Catal. 215 (2003) 353.
56. H. Farag, I. Mochida, K. Sakanishi, Appl. Catal. A 194-195 (2000) 147.
57. J.A.R. van Veen, H.A. Colijn, P.A.J.M. Hendriks, A.J. van Welsenes, Fuel. Proc. Technol. 35 (1993) 137.
58. E. J. M. Hensen, M. J. Vissenberg, V. H. J. de Beer, J. A. R. van Veen, R. A. van Santen, J. Catal. 163 (1996) 429.
59. E. J. M. Hensen, Ph.D. Thesis, Eindhoven University of Technology, Eindhoven, 2000, ISBN 90-386-2871-4.

Chapter 6

Formation of active phases in $CoMo/Al_2O_3$ catalysts prepared using NTA and phosphate

Abstract

The influence of the sulfiding pressure on the structure and activity of γ-alumina supported catalysts prepared with nitrilotriacetic acid (NTA) and phosphorus was studied by Mössbauer emission spectroscopy (MES), extended X-ray absorption fine structure (EXAFS), transmission electron microscopy (TEM) and dibenzothiophene hydrodesulfurization (HDS) activity measurements. The presence of additives that decrease the metal-support interactions leads to a significant formation of CoS_x species involving variable number of Co atoms in the high promoter-loading catalysts. A more significant presence of the CoS_x structures was observed with the phosphorus-containing catalyst, where also the weak interaction of Mo with the support led to higher loss in the MoS_2 dispersion as compared to the similar sample prepared without phosphate. The spectroscopic data of the stepwise sulfidation approach indicate a complete formation of Type II Co-Mo-S species only after the direct high-pressure treatment.

6.1 Introduction

It is widely accepted that the most important factors that determine the catalytic activity of supported Co(Ni)Mo sulfides are the dispersion and morphology of the active phase [1]. Two types of Co-Mo-S phases have been introduced, a Type II in which all Mo–O–Al linkages with the support are completely sulfided [2] with a two times higher activity in gas-phase desulfurization than the not fully sulfided Type I counterpart [3-5].

The use of complexing agents, which decrease Mo-support interactions [6,7], can lead to Type II Co-Mo-S phase formation. In the same time, the complexing agents retard sulfidation of the Co atoms until complete Mo sulfidation [8,9] favoring better promotion of MoS_2 by Co. Moreover, the decreased Mo-support interaction due to the presence of complexing agents like NTA (nitrilotriacetic acid) during sulfidation leads to a more stacked morphology. It was also shown [8] that the chelating agents preferentially form complexes with Ni, with molybdate species only being complexed after all Ni ions are complexed to the organic molecule.

For catalysts activated under atmospheric pressure conditions, it was found that Type II Co(Ni)-Mo-S is much more active in gas-phase HDS reactions than Type I, but slightly less active in trickle-flow dibenzothiophene (DBT) HDS [10].

Literature [11,12] indicates that introduction of phosphorus to molybdenum-based catalysts improves the thiophene HDS activity. Phosphorus is usually added as phosphoric acid to the alumina support leading to the formation of an aluminum phosphate surface phase that is stable during treatments and hydrotreating operations below 500 ºC [13]. Different explanations for the promoting role of phosphorus were proposed: reduced amounts of inactive Co and Ni species impeded by the phosphorus interaction with the support [14], improved dispersion [15,16] and increased stacking of MoS_2 layers [9,17].

To better understand the sulfidation mechanism of $CoMo/Al_2O_3$ catalysts prepared via these non-conventional routes also at higher pressures, we investigated NTA and phosphorus containing CoMo catalysts. The formation and evolution of the active phase as a function of sulfidation treatment were followed using (Mössbauer Emission Spectroscopy - MES) and EXAFS (Mo and Co edges). The characterization of the catalysts was completed with high-resolution TEM investigations of the morphology and dispersion of the active structures and catalytic activity measurements (DBT HDS).

6.2 Experimental

Seven CoMo catalysts were prepared by pore-volume impregnation using γ-Al_2O_3 (Ketjen 001-1.5E, BET surface area 271 $m^2\ g^{-1}$, pore volume 0.7 ml g^{-1}, particle size 0.5-0.85 mm) as support material. Ammoniacal solution containing ammonium heptamolybdate $(NH_4)_6Mo_7O_{24}\cdot 4H_2O$ (Merck, min. 99.9%), cobalt nitrate $Co(NO_3)_2\cdot 6H_2O$ (Merck p.a.) and NTA was used in a one-step impregnation procedure as described by Van Veen *et al.* [3]. One phosphorus-containing catalyst was prepared using an aqueous solution of MoO_3, H_3PO_4 and cobalt salts [3,18] (half of the Co was introduced as $CoCO_3$ and half as $Co(NO_3)_2\cdot 6H_2O$ because of the low solubility of $CoCO_3$). About 50 MBq ^{57}Co as an aqueous solution of $Co(NO_3)_2\cdot 6H_2O$ was added to the impregnation solutions, for the MES measurements.

The catalysts contained 7 wt% Mo, with four high promoter-loading catalysts having 2.25 wt% Co, denoted as Co(2.25)Mo(7)/Al_2O_3, and two catalysts having 1.3 wt% Co, denoted as Co(1.3)Mo(7)/Al_2O_3. The employed NTA:Co and NTA:Mo molar ratios are also indicated in the label of the catalysts. The phosphorus catalyst contained 1.18 wt% P and 2.25 wt% Co and is denoted as Co(2.25)Mo(7)P(1.18)/Al_2O_3. The catalyst loadings are given relative to the support material and are calculated from the impregnation solutions. After impregnation the catalysts were dried in static air at 383 K for 16 h. The Co(2.25)Mo(7)P(1.18)/Al_2O_3 catalyst was subsequently heated up to 673 K in 5 h and calcined at this temperature for 24 h in static air. The applied sulfidation treatment is denoted (S, x MPa, y K), indicating that during the experiment the catalyst is linearly heated to y K at x MPa (heating rate 2 °C/min) and kept at that temperature for 1 h. The phosphorus-containing catalyst was linearly heated to the treatment temperature in 1 h. For the NTA-containing catalysts, the calcination treatment was omitted and a low heating rate was used in order to avoid the premature decomposition of the metal:NTA complexes.

The experimental method applied for the MES measurements is presented in section 2.1.4 and the details of the EXAFS measurements and analysis are available in 2.2.2. For the TEM experiments described in section 2.3, the same sulfided samples were used as for the EXAFS measurements. Gas-phase DBT HDS experiments were carried out in a high-pressure stainless steel reactor according to the method described in 2.4. The rate constants (k_{HDS}) measured after 8 hours at 3 MPa and 573 K are expressed per kg of Mo.

6.3 Results

6.3.1 MES results

The Mössbauer spectra of the Co(2.25)Mo(7)/Al_2O_3 NTA:Mo=1.2 catalyst sulfided at 4 MPa are presented in Figure 6.1 and the resulting MES parameters are shown in Table 6.1. A high-spin 2+ contribution is observed largely in the MES spectrum of the fresh catalyst, indicating that the Co:NTA complex is still intact after the drying treatment [3,5]. After exposing the catalyst to the H_2S/H_2 gas mixture at room temperature and atmospheric pressure, the high-spin 2+ phase contribution increases. This behavior is similar to that observed with catalysts prepared without NTA.

The high-spin 2+ contribution completely disappears from the spectrum after treatment at 373 K and 4 MPa. The Q.S. of the second doublet is continuously decreasing after sulfidation at 373 K and only changes little after sulfidation at 573 K. After activation at 673 K and 4 MPa, the resulting doublet has an IS of 0.22 mm s^{-1} and a QS of 0.86 mm s^{-1}.

Figure 6.2 shows the MES spectra for Co(2.25)Mo(7)/Al_2O_3 NTA:Mo=1.2 catalyst sulfided at atmospheric pressure and then further sulfided at 4 MPa. The corresponding fit parameters are presented in Table 6.2. The high-spin 2+ contribution increases also after sulfidation at 373 K, disappearing from the spectrum only after treatment at 573 K and 0.1 MPa. The Q.S. values of the second spectral contribution continuously decrease and are smaller than the values obtained with the catalyst sulfided at high-pressure. A significant decrease of QS is observed when the sulfidation temperature is raised from 473 to 573 K.

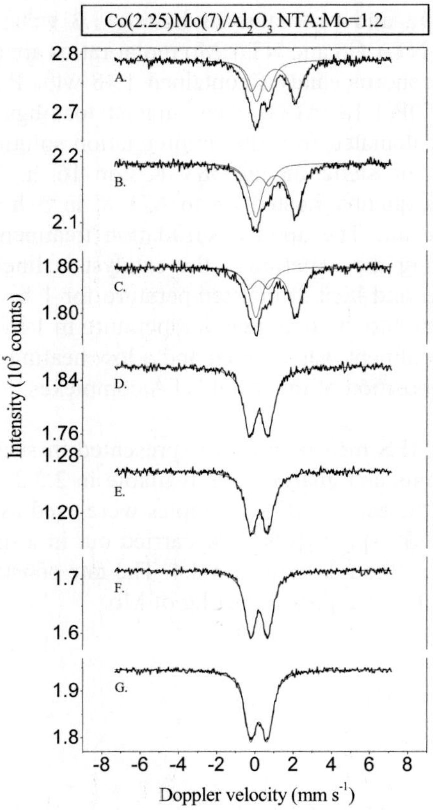

Figure 6.1 MES spectra obtained at 300 K with Co(2.25)Mo(7)/Al_2O_3 NTA:Mo=1.2 catalyst after various successive sulfidation steps.
A. Fresh catalyst;
B. (S, 0.1 MPa, 300 K);
C. (S, 4 MPa, 300 K);
D. (S, 4 MPa, 373 K);
E. (S, 4 MPa, 473 K);
F. (S, 4 MPa, 573 K);
G. (S, 4 MPa, 673 K).

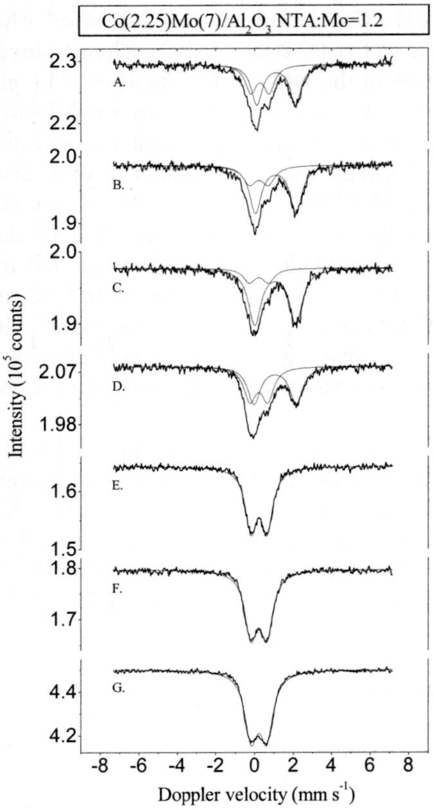

Figure 6.2 MES spectra obtained at 300 K with Co(2.25)Mo(7)/Al_2O_3 NTA:Mo=1.2 catalyst after various successive sulfidation steps.
A. Fresh catalyst;
B. (S, 0.1 MPa, 300 K);
C. (S, 0.1 MPa, 373 K);
D. (S, 0.1 MPa, 473 K);
E. (S, 0.1 MPa, 573 K);
F. (S, 0.1 MPa, 673 K);
G. (S, 4 MPa, 673 K).

Table 6.1 *MES parameters of $Co(2.25)Mo(7)/Al_2O_3$ NTA:Mo=1.2 catalyst after sulfidation treatment*

T_s K	P MPa	IS $mm \cdot s^{-1}$	QS $mm \cdot s^{-1}$	Γ $mm \cdot s^{-1}$	A %	IS $mm \cdot s^{-1}$	QS $mm \cdot s^{-1}$	Γ $mm \cdot s^{-1}$	A %	IS $mm \cdot s^{-1}$	QS $mm \cdot s^{-1}$	Γ $mm \cdot s^{-1}$	A %
		"Co – oxide"				"Co – sulfide"				"High-spin 2+"			
Fresh		0.28	0.91	0.69	39					1.12	2.05	0.75	61
300	0.1					0.23	1.04	0.74	27	1.09	2.09	0.77	73
300	4					0.19	1.02	0.75	29	1.09	2.08	0.73	71
373	4					0.21	0.96	0.74	100				
473	4					0.23	0.89	0.71	100				
573	4					0.23	0.90	0.66	100				
673	4					0.22	0.86	0.73	100				

Table 6.2 *MES parameters of $Co(2.25)Mo(7)/Al_2O_3$ NTA:Mo=1.2 catalyst after sulfidation treatment*

T_s K	P MPa	IS $mm \cdot s^{-1}$	QS $mm \cdot s^{-1}$	Γ $mm \cdot s^{-1}$	A %	IS $mm \cdot s^{-1}$	QS $mm \cdot s^{-1}$	Γ $mm \cdot s^{-1}$	A %	IS $mm \cdot s^{-1}$	QS $mm \cdot s^{-1}$	Γ $mm \cdot s^{-1}$	A %
		"Co – oxide"				"Co – sulfide"				"High-spin 2+"			
Fresh		0.28	0.95	0.69	40					1.12	2.01	0.70	60
300	0.1					0.23	0.97	0.84	30	1.10	2.10	0.73	70
373	0.1					0.22	0.99	0.73	19	1.07	2.13	0.79	81
473	0.1					0.22	0.92	0.76	46	1.06	2.17	0.79	54
573	0.1					0.23	0.82	0.72	100				
673	0.1					0.22	0.81	0.76	100				
673	4					0.23	0.80	0.80	100				

The comparative evolution of the Q.S. values of the Co-sulfide doublet after sulfidation of Co(2.25)Mo(7)/Al_2O_3 NTA:Mo=1.2 at atmospheric and high-pressure is shown in Figure 6.3.

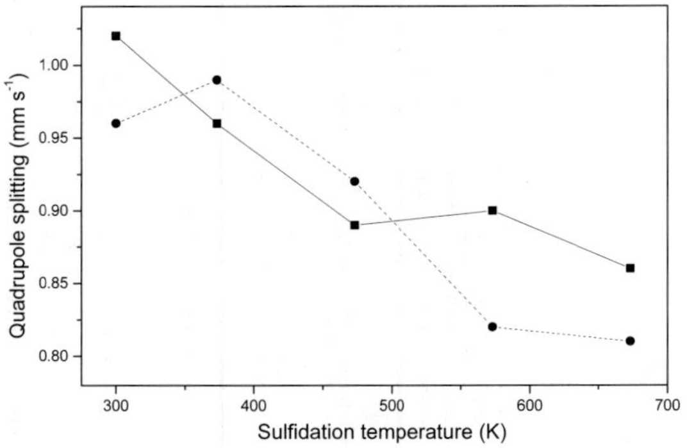

Figure 6.3 *Q.S. values of the Co-sulfide doublet obtained with Co(2.25)Mo(7)/Al_2O_3 NTA:Mo=1.2 catalyst at 0.1 MPa (●) and at 4 MPa (■) after stepwise sulfidation at the indicated temperatures.*

The Mössbauer spectra for a catalyst containing a lower amount of the NTA complexing agent (Co(2.25)Mo(7)/Al_2O_3 NTA:Co=1) after sulfidation at 4 MPa are presented in Figure 6.4 and the resulting MES parameters are collected in Table 6.3.

The high-spin 2+ spectral contribution in the fresh sample is smaller compared to that of the catalyst containing more NTA. This contribution also disappears from the spectrum after treatment at 373 K and 4 MPa. The Q.S. of the remaining doublet increases slightly after sulfidation at 573 K, a doublet with QS of 0.84 mm s^{-1} being found upon treatment at 673 K and 4 MPa.

Figure 6.5 shows the MES spectra of Co(2.25)Mo(7)/Al_2O_3 NTA:Co=1 sulfided at atmospheric pressure and subsequently at high-pressure. The corresponding fit parameters are presented in Table 6.4. The sulfidation behavior of this catalyst is similar to that of the previous sample sulfided first at atmospheric pressure. The high-spin 2+ contribution disappears from the spectra only after treatment at 573 K and 0.1 MPa and the Q.S. values of the second spectral contribution continuously decrease, also after increasing the sulfidation temperature from 473 K to 573 K. Upon atmospheric and high-pressure activation at 673 K, a small doublet with an IS of 0.23 mm s^{-1} and a QS of 0.77 mm s^{-1} is observed in the spectrum. The evolution of the Q.S. values for the Co(2.25)Mo(7)/Al_2O_3 NTA:Co=1 catalyst upon treatment under atmospheric and high-pressure conditions is presented in Figure 6.6.

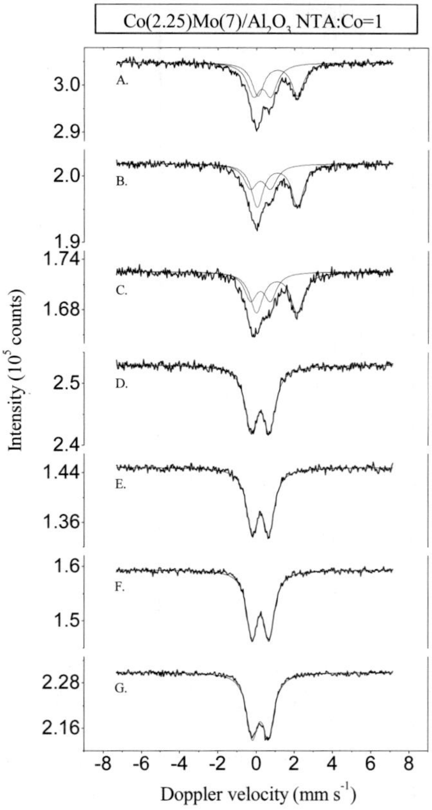

Figure 6.4 MES spectra obtained at 300 K with Co(2.25)Mo(7)/Al_2O_3 NTA:Co=1 catalyst after various successive sulfidation steps.

A. *Fresh catalyst;*
B. *(S, 0.1 MPa, 300 K);*
C. *(S, 4 MPa, 300 K);*
D. *(S, 4 MPa, 373 K);*
E. *(S, 4 MPa, 473 K);*
F. *(S, 4 MPa, 573 K);*
G. *(S, 4 MPa, 673 K).*

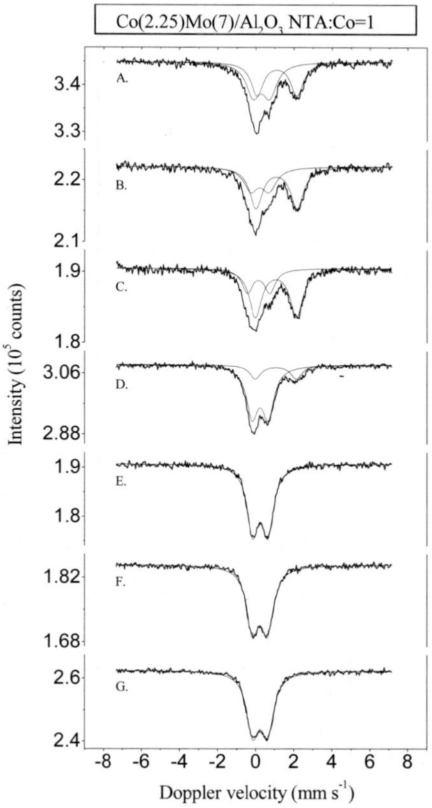

Figure 6.5 MES spectra obtained at 300 K with Co(2.25)Mo(7)/Al_2O_3 NTA:Co=1 catalyst after various successive sulfidation steps.

A. *Fresh catalyst;*
B. *(S, 0.1 MPa, 300 K);*
C. *(S, 0.1 MPa, 373 K);*
D. *(S, 0.1 MPa, 473 K);*
E. *(S, 0.1 MPa, 573 K);*
F. *(S, 0.1 MPa, 673 K);*
G. *(S, 4 MPa, 673 K).*

Table 6.3 *MES parameters of Co(2.25)Mo(7)/Al_2O_3 NTA:Co=1 catalyst after sulfidation treatment*

T_s K	P MPa	IS $mm \cdot s^{-1}$	QS $mm \cdot s^{-1}$	Γ $mm \cdot s^{-1}$	A %	IS $mm \cdot s^{-1}$	QS $mm \cdot s^{-1}$	Γ $mm \cdot s^{-1}$	A %	IS $mm \cdot s^{-1}$	QS $mm \cdot s^{-1}$	Γ $mm \cdot s^{-1}$	A %
		“Co – oxide”				“Co – sulfide”				“High-spin 2+”			
Fresh		0.30	0.88	0.90	51					1.10	2.10	0.76	49
300	0.1					0.21	1.05	0.83	37	1.10	2.12	0.75	63
300	4					0.21	1.03	0.79	39	1.07	2.14	0.82	61
373	4					0.21	0.95	0.75	100				
473	4					0.22	0.86	0.68	100				
573	4					0.22	0.89	0.66	100				
673	4					0.22	0.84	0.71	100				

Table 6.4 *MES parameters of Co(2.25)Mo(7)/Al_2O_3 NTA:Co=1 catalyst after sulfidation treatment*

T_s K	P MPa	IS $mm \cdot s^{-1}$	QS $mm \cdot s^{-1}$	Γ $mm \cdot s^{-1}$	A %	IS $mm \cdot s^{-1}$	QS $mm \cdot s^{-1}$	Γ $mm \cdot s^{-1}$	A %	IS $mm \cdot s^{-1}$	QS $mm \cdot s^{-1}$	Γ $mm \cdot s^{-1}$	A %
		“Co – oxide”				“Co – sulfide”				“High-spin 2+”			
Fresh		0.28	0.88	0.90	52					1.11	2.09	0.80	48
300	0.1					0.21	0.96	0.93	38	1.08	2.16	0.83	62
373	0.1					0.22	1.18	0.75	31	1.06	2.17	0.81	69
473	0.1					0.24	0.84	0.75	79	1.06	2.16	0.72	21
573	0.1					0.24	0.79	0.72	100				
673	0.1					0.23	0.77	0.77	100				
673	4					0.23	0.77	0.79	100				

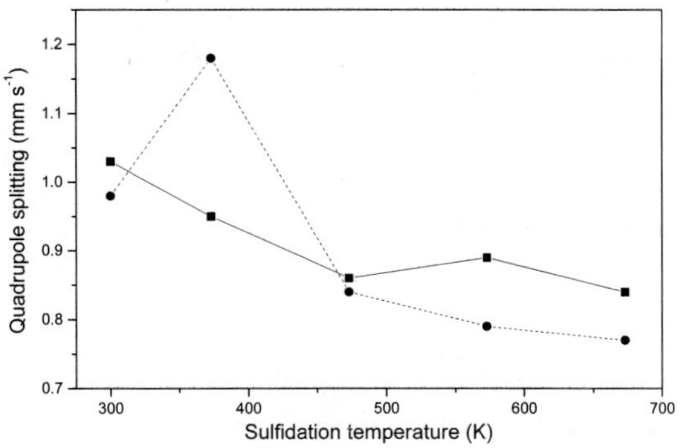

Figure 6.6 *Q.S. values of the Co-sulfide doublet obtained with $Co(2.25)Mo(7)/Al_2O_3$ NTA:Co=1 catalyst at 0.1 MPa (●) and at 4 MPa (■) after stepwise sulfidation at the indicated temperatures.*

The Mössbauer spectra of the $Co(1.3)Mo(7)/Al_2O_3$ NTA:Mo=1.2 catalyst sulfided at atmospheric pressure and then subjected to a high-pressure treatment are presented in Figure 6.7 and the resulting MES parameters are shown in Table 6.5.

After treatment at 573 K and 0.1 MPa, the high-spin 2+ contribution completely disappears from the spectrum and a significant increase in the Q.S. value of the remaining doublet is observed. Upon atmospheric and high-pressure sulfidation at 673 K, a doublet with an IS of 0.22 mm s^{-1} and a QS of 1.08 mm s^{-1} is present in the spectra.

Figure 6.8 shows the MES spectra obtained with the $Co(1.3)Mo(7)/Al_2O_3$ NTA:Co=1 catalyst sulfided at atmospheric pressure and subsequently at high-pressure and the corresponding fit parameters are presented in Table 6.6. The high-spin 2+ contribution is smaller after atmospheric pressure treatments at 373 K and 473 K compared to that of the similar sample containing more NTA. This contribution again disappears from the spectrum after treatment at 573 K and 0.1 MPa. Once more, the Q.S. of the remaining doublet increases upon activation at 573 K.

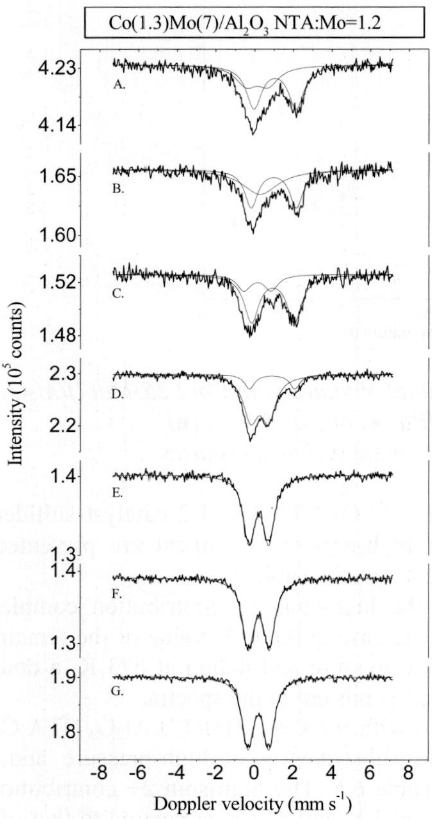

Figure 6.7 MES spectra obtained at 300 K with Co(1.3)Mo(7)/Al$_2$O$_3$ NTA:Mo=1.2 catalyst after various successive sulfidation steps.
A. Fresh catalyst;
B. (S, 0.1 MPa, 300 K);
C. (S, 0.1 MPa, 373 K);
D. (S, 0.1 MPa, 473 K);
E. (S, 0.1 MPa, 573 K);
F. (S, 0.1 MPa, 673 K);
G. (S, 4 MPa, 673 K).

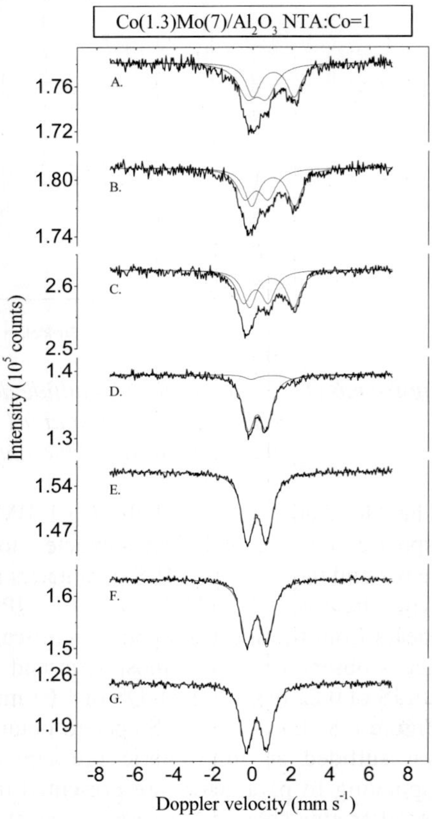

Figure 6.8 MES spectra obtained at 300 K with Co(1.3)Mo(7)/Al$_2$O$_3$ NTA:Co=1 catalyst after various successive sulfidation steps.
A. Fresh catalyst;
B. (S, 0.1 MPa, 300 K);
C. (S, 0.1 MPa, 373 K);
D. (S, 0.1 MPa, 473 K);
E. (S, 0.1 MPa, 573 K);
F. (S, 0.1 MPa, 673 K);
G. (S, 4 MPa, 673 K).

Table 6.5 *MES parameters of Co(1.3)Mo(7)/Al_2O_3 NTA:Mo=1.2 catalyst after sulfidation treatment*

T_s K	P MPa	IS $mm{\cdot}s^{-1}$	QS $mm{\cdot}s^{-1}$	Γ $mm{\cdot}s^{-1}$	A %	IS $mm{\cdot}s^{-1}$	QS $mm{\cdot}s^{-1}$	Γ $mm{\cdot}s^{-1}$	A %	IS $mm{\cdot}s^{-1}$	QS $mm{\cdot}s^{-1}$	Γ $mm{\cdot}s^{-1}$	A %
		"Co – oxide"				"Co – sulfide"				"High-spin 2+"			
Fresh		0.20	1.14	1.34	38					1.05	2.18	0.96	62
300	0.1					0.18	1.40	1.53	42	1.04	2.24	0.80	58
373	0.1					0.15	1.41	0.85	24	0.94	2.21	0.92	76
473	0.1					0.26	0.92	0.89	81	0.91	2.34	0.63	19
573	0.1					0.22	1.02	0.70	100				
673	0.1					0.22	1.08	0.72	100				
673	4					0.22	1.08	0.69	100				

Table 6.6 *MES parameters of Co(1.3)Mo(7)/Al_2O_3 NTA:Co=1 catalyst after sulfidation treatment*

T_s K	P MPa	IS $mm{\cdot}s^{-1}$	QS $mm{\cdot}s^{-1}$	Γ $mm{\cdot}s^{-1}$	A %	IS $mm{\cdot}s^{-1}$	QS $mm{\cdot}s^{-1}$	Γ $mm{\cdot}s^{-1}$	A %	IS $mm{\cdot}s^{-1}$	QS $mm{\cdot}s^{-1}$	Γ $mm{\cdot}s^{-1}$	A %
		"Co – oxide"				"Co – sulfide"				"High-spin 2+"			
Fresh		0.20	0.97	1.16	53					1.06	2.10	0.89	47
300	0.1					0.19	1.17	0.99	46	1.04	2.19	0.86	54
373	0.1					0.16	1.25	0.87	46	0.96	2.25	0.84	54
473	0.1					0.24	0.90	0.73	91	0.99	2.14	0.81	9
573	0.1					0.22	0.98	0.72	100				
673	0.1					0.22	1.03	0.71	100				
673	4					0.22	1.04	0.70	100				

Table 6.7 *MES parameters of Co(2.25)Mo(7)P(1.18)/Al_2O_3 catalyst after sulfidation treatment*

T_s K	P MPa	IS $mm{\cdot}s^{-1}$	QS $mm{\cdot}s^{-1}$	Γ $mm{\cdot}s^{-1}$	A %	IS $mm{\cdot}s^{-1}$	QS $mm{\cdot}s^{-1}$	Γ $mm{\cdot}s^{-1}$	A %	IS $mm{\cdot}s^{-1}$	QS $mm{\cdot}s^{-1}$	Γ $mm{\cdot}s^{-1}$	A %
		"Co – oxide"				"Co – sulfide"				"High-spin 2+"			
Fresh		0.29	0.89	0.89	69					0.81	2.43	0.81	31
300	0.1					0.21	1.05	0.83	56	0.94	2.18	0.80	44
300	4					0.24	0.96	0.86	57	0.92	2.22	0.78	43
373	4					0.23	0.87	0.75	71	0.91	2.13	0.79	29
473	4					0.23	0.85	0.69	86	1.06	2.00	0.75	14
573	4					0.24	0.85	0.70	94	1.09	1.78	0.60	6
673	4					0.24	0.75	0.82	100				

The Mössbauer spectra of the Co(2.25)Mo(7)P(1.18)/Al_2O_3 catalyst sulfided at 4 MPa are presented in Figure 6.9 and the resulting MES parameters are shown in Table 6.7.

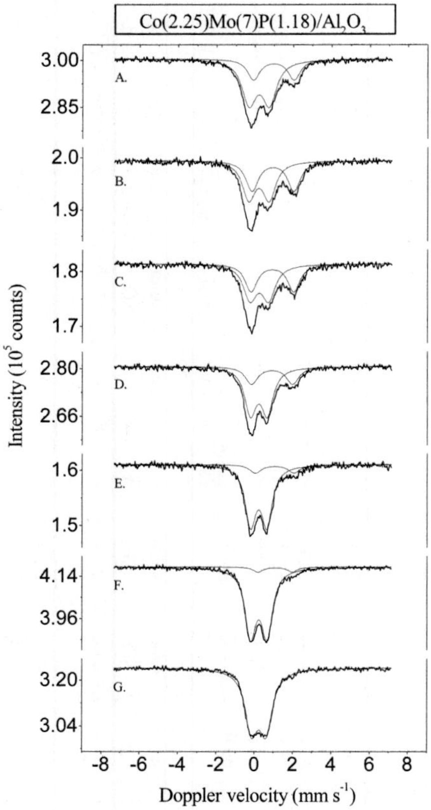

Figure 6.9 *MES spectra obtained at 300 K with Co(2.25)Mo(7)P(1.18)/Al_2O_3 catalyst after various successive sulfidation steps. A. Fresh catalyst; B. (S, 0.1 MPa, 300 K); C. (S, 4 MPa, 300 K); D. (S, 4 MPa, 373 K); E. (S, 4 MPa, 473 K); F. (S, 4 MPa, 573 K); G. (S, 4 MPa, 673 K).*

The high-spin 2+ contribution disappears from the spectra only after treatment at 673 K and 4 MPa. This behavior also observed in the previous MES study of the similar calcined catalyst prepared without phosphorus [19]. The Q.S. values of the second spectral contribution continuously decrease, a doublet with QS of 0.75 mm s^{-1} being measured upon sulfidation at 673 K and 4 MPa.

6.3.2 Mo K-edge EXAFS

The Mo K-edge EXAFS parameters of Co(2.25)Mo(7)/Al_2O_3 NTA:Mo=1.2 sulfided at intermediate temperatures under atmospheric and high-pressure conditions are collected in Table 6.8. The spectrum of the sample sulfided at 473 K and 4 MPa was fitted with a Mo-S contribution at 2.38 Å and a Mo-Mo contribution at 2.83 Å. The Mo-S and Mo-Mo distances are smaller than the corresponding distances in crystalline MoS_2 [20]. An overlapping of a Mo-O contribution and the Mo-S shell or a different Mo valence can be suggested by the high Mo-S coordination number observed at this temperature [21,22].

Table 6.8 *Mo K-edge EXAFS parameters of Co(2.25)Mo(7)/Al_2O_3 NTA:Mo=1.2 catalyst after different sulfidation treatments*

Treatment	N -	R Å	$\Delta\sigma^2$ Å^2 $\cdot 10^{-3}$	ΔE_0 eV	N -	R Å	$\Delta\sigma^2$ Å^2 $\cdot 10^{-3}$	ΔE_0 eV	N -	R Å	$\Delta\sigma^2$ Å^2 $\cdot 10^{-3}$	ΔE_0 eV
	Mo-S				Mo-Mo				Mo-Co			
S, 4 MPa, 473 K	5.8	2.38	8.1	6.7	2.5	2.83	9.5	-3.0				
S, 4 MPa, 573 K	5.7	2.41	2.5	2.9	2.8	3.15	5.2	5.0	0.9	2.77	9.2	9.9
S, 4 MPa, 673 K	6.1	2.42	1.6	1.7	4.0	3.16	4.0	2.3	0.7	2.76	6.9	9.1
S, 0.1 MPa, 473 K	5.3	2.39	8.3	3.1	1.7	2.82	9.2	-5.7				
S, 0.1 MPa, 573 K	5.5	2.41	3.1	1.8	2.7	3.15	5.3	4.8	0.5	2.77	9.0	9.5
S, 0.1 MPa, 673 K	5.8	2.41	1.6	0.8	3.8	3.16	3.8	1.8	1.0	2.75	9.5	9.8
S, 0.1 MPa + S, 4 MPa, 673 K	5.9	2.41	1.5	1.2	3.8	3.16	3.5	2.2	1.0	2.75	9.6	9.5

Table 6.9 *Mo K-edge EXAFS parameters of Co(2.25)Mo(7)/Al_2O_3 NTA:Co=1 catalyst after different sulfidation treatments*

Treatment	N -	R Å	$\Delta\sigma^2$ Å^2 $\cdot 10^{-3}$	ΔE_0 eV	N -	R Å	$\Delta\sigma^2$ Å^2 $\cdot 10^{-3}$	ΔE_0 eV	N -	R Å	$\Delta\sigma^2$ Å^2 $\cdot 10^{-3}$	ΔE_0 eV
	Mo-S				Mo-Mo				Mo-Co			
S, 4 MPa, 473 K	5.5	2.37	8.0	8.4	2.5	2.83	8.0	-1.4				
S, 4 MPa, 573 K	5.5	2.41	2.5	1.8	2.8	3.16	5.4	1.1	0.7	2.75	9.5	9.9
S, 4 MPa, 673 K	5.9	2.41	1.3	1.9	4.0	3.16	3.7	2.9	1.1	2.75	9.5	9.6
S, 0.1 MPa, 473 K	5.4	2.39	7.7	3.5	1.6	2.83	8.6	-7.2				
S, 0.1 MPa, 573 K	5.5	2.41	3.1	1.3	2.6	3.15	5.4	3.6	0.6	2.76	9.7	9.4
S, 0.1 MPa, 673 K	5.7	2.41	1.4	1.8	3.8	3.16	3.9	2.8	0.9	2.75	9.6	9.8
S, 0.1 MPa + S, 4 MPa, 673 K	5.7	2.41	1.3	1.8	3.7	3.16	3.7	1.6	0.8	2.75	9.6	7.6

After increasing the sulfidation temperature to 573 K and 673 K, the Mo-S and Mo-Mo interatomic distances correspond to the structural parameters of bulk MoS_2, while the observed Mo-Mo coordination number is lower than that of MoS_2 ($N_{Mo-Mo} = 6$). A Mo-Co contribution is fitted with a coordination distance of about 2.77 Å after activation at 573 K and 4 MPa. This contribution does not change significantly after the high-pressure treatment at 673 K. A similar behavior is observed with the sample sulfided first at atmospheric pressure. The N_{Mo-Co} increases after treatment at 673 K and 0.1 MPa as compared to 573 K and remains constant after increasing the sulfidation pressure.

The fit results of the Mo K-edge EXAFS data of Co(2.25)Mo(7)/Al_2O_3 NTA:Co=1 sulfided at 0.1 MPa and 4 MPa are shown in Table 6.9. Again, the Mo-S and Mo-Mo distances are smaller than the corresponding distances in the crystalline MoS_2 after activation at 473 K and different pressures, but upon increasing the sulfidation temperature the presence of MoS_2-like species is evident at all intermediate steps. The Mo-Co contribution is also present after treatments at 573 K and 673 K with similar structural parameters as found for the samples with a higher NTA loading.

The Mo K-edge EXAFS parameters of the Co(2.25)Mo(7)P(1.18)/Al_2O_3 catalyst sulfided under high-pressure conditions are given in Table 6.10. MoS_2-like species are present already after activation at 473 K and 4 MPa. Relatively small Mo-S coordination numbers are found compared to the catalysts prepared with NTA indicating that some of the Mo edge atoms are still coordinated to oxygen or that the MoS_2 edges are not completely saturated with sulfur [23]. The Mo-Co contribution is found at a distance of 2.75 Å after increasing the temperature to 573 K and a somewhat lower Mo-Mo coordination number as compared to the NTA catalysts is fitted after treatment at 673 K.

Table 6.10 *Mo K-edge EXAFS parameters of Co(2.25)Mo(7)P(1.18)/Al_2O_3 catalyst after different sulfidation treatments*

Treatment	N -	R Å	$\Delta\sigma^2$ $Å^2$ $\cdot10^{-3}$	ΔE_0 eV	N -	R Å	$\Delta\sigma^2$ $Å^2$ $\cdot10^{-3}$	ΔE_0 eV	N -	R Å	$\Delta\sigma^2$ $Å^2$ $\cdot10^{-3}$	ΔE_0 eV
	Mo-S				Mo-Mo				Mo-Co			
S, 4 MPa, 473 K	4.7	2.41	6.3	2.9	1.7	3.11	9.7	9.9				
S, 4 MPa, 573 K	4.8	2.41	3.4	-1.1	2.3	3.15	6.2	4.2	0.9	2.75	9.8	9.7
S, 4 MPa, 673 K	5.4	2.41	1.5	0.2	3.4	3.16	3.8	1.7	1.1	2.74	9.6	9.6

6.3.3 Co K-edge EXAFS

The Co K-edge EXAFS fit parameters of the sulfided Co(2.25)Mo(7)/Al_2O_3 NTA:Mo=1.2 catalyst are listed in Table 6.11. The fit parameters are similar for the samples activated under atmospheric and high-pressure conditions. The spectra were fitted with a Co-S contribution at a distance of 2.23 Å and two Co-Co contributions at 2.6 Å and 3.33 Å. These distances are slightly different from those corresponding in bulk Co_9S_8 (2.51 Å and 3.51 Å, respectively) [24]. A Co-Mo contribution at 2.76 Å could be fitted in the spectra after treatment at 573 K and 673 K and a somewhat smaller $N_{Co-Co(1)}$ is obtained for the sample sulfided directly at high-pressure.

Table 6.11 *Co K-edge EXAFS parameters of Co(2.25)Mo(7)/Al$_2$O$_3$ NTA:Mo=1.2 catalyst after different sulfidation treatments*

Treatment	N -	R Å	$\Delta\sigma^2$ Å^2 $\cdot10^{-3}$	ΔE_0 eV	N -	R Å	$\Delta\sigma^2$ Å^2 $\cdot10^{-3}$	ΔE_0 eV	N -	R Å	$\Delta\sigma^2$ Å^2 $\cdot10^{-3}$	ΔE_0 eV	N -	R Å	$\Delta\sigma^2$ Å^2 $\cdot10^{-3}$	ΔE_0 eV
	Co-S				Co-Co (1)				Co-Mo				Co-Co (2)			
S, 4 MPa, 473 K	5.7	2.23	3.9	-1.1	1.0	2.56	9.1	-3.1					1.6	3.29	9.8	9.5
S, 4 MPa, 573 K	5.6	2.23	3.6	-1.3	0.9	2.60	8.1	-2.3	0.7	2.76	9.2	-9.6	0.5	3.32	2.0	7.2
S, 4 MPa, 673 K	5.7	2.23	3.8	-0.9	0.7	2.60	7.2	-5.8	0.8	2.76	9.4	-9.8	0.5	3.33	0.1	7.5
S, 0.1 MPa, 473 K	5.7	2.23	3.9	-0.9	1.0	2.57	9.4	-4.6					1.5	3.28	9.7	9.4
S, 0.1 MPa, 573 K	5.6	2.23	3.9	-1.0	1.0	2.60	9.8	-4.9	0.5	2.76	9.9	-9.4	0.7	3.33	3.5	5.7
S, 0.1 MPa, 673 K	5.6	2.23	3.7	-1.0	0.9	2.60	9.6	-5.9	0.5	2.76	9.9	-9.8	0.6	3.32	2.6	7.0
S, 0.1 MPa + S, 4 MPa, 673 K	5.6	2.23	3.6	-1.1	1.0	2.60	9.5	-3.5	0.8	2.76	9.8	-9.2	0.5	3.33	1.7	6.8

Table 6.12 *Co K-edge EXAFS parameters of Co(2.25)Mo(7)/Al_2O_3 NTA:Co=1 catalyst after different sulfidation treatments*

Treatment	N -	R Å	$\Delta\sigma^2$ Å^2 $\cdot 10^{-3}$	ΔE_0 eV	N -	R Å	$\Delta\sigma^2$ Å^2 $\cdot 10^{-3}$	ΔE_0 eV	N -	R Å	$\Delta\sigma^2$ Å^2 $\cdot 10^{-3}$	ΔE_0 eV	N -	R Å	$\Delta\sigma^2$ Å^2 $\cdot 10^{-3}$	ΔE_0 eV
	Co-S				Co-Co (1)				Co-Mo				Co-Co (2)			
S, 4 MPa, 473 K	5.6	2.23	3.7	-1.1	1.2	2.56	9.5	-1.0					1.6	3.28	9.7	9.6
S, 4 MPa, 573 K	5.5	2.23	3.7	-1.1	1.0	2.60	9.5	-3.5	0.7	2.76	9.9	-9.8	0.5	3.33	2.2	6.6
S, 4 MPa, 673 K	5.6	2.23	3.6	-1.2	1.0	2.60	9.1	-3.2	0.8	2.76	9.9	-9.2	0.5	3.33	1.7	6.8
S, 0.1 MPa, 473 K	5.5	2.23	4.4	-0.9	1.3	2.59	6.4	-6.3					1.2	3.28	9.9	9.5
S, 0.1 MPa, 573 K	5.6	2.23	4.1	-1.0	1.1	2.59	9.8	-1.2	0.5	2.76	4.2	-0.9	1.1	3.32	9.8	9.7
S, 0.1 MPa, 673 K	5.6	2.23	4.0	-1.2	1.1	2.60	9.7	-2.2	0.5	2.76	2.3	-2.1	1.1	3.33	9.7	9.6
S, 0.1 MPa + S, 4 MPa, 673 K	5.5	2.23	3.6	-1.1	1.2	2.60	9.1	-3.9	0.7	2.76	9.7	-9.5	0.8	3.32	3.9	9.6

Table 6.13 *Co K-edge EXAFS parameters of Co(2.25)Mo(7)P(1.18)/Al_2O_3 catalyst after different sulfidation treatments*

Treatment	N -	R Å	$\Delta\sigma^2$ Å^2 $\cdot 10^{-3}$	ΔE_0 eV	N -	R Å	$\Delta\sigma^2$ Å^2 $\cdot 10^{-3}$	ΔE_0 eV	N -	R Å	$\Delta\sigma^2$ Å^2 $\cdot 10^{-3}$	ΔE_0 eV	N -	R Å	$\Delta\sigma^2$ Å^2 $\cdot 10^{-3}$	ΔE_0 eV
	Co-S				Co-Co (1)				Co-Mo				Co-Co (2)			
S, 4 MPa, 473 K	5.6	2.22	4.3	1.0	1.1	2.59	6.2	-9.6					0.8	3.48	9.5	-8.6
S, 4 MPa, 573 K	5.9	2.22	4.5	0.1	0.5	2.56	9.9	-5.7	1.0	2.80	7.6	-4.7	0.6	3.48	9.3	-9.7
S, 4 MPa, 673 K	6.0	2.22	4.7	0.3	0.4	2.56	9.4	-8.2	1.1	2.80	7.8	-2.9	0.7	3.48	9.8	-9.5

The Co K-edge fit results for the sample with a lower NTA loading (Table 6.12) show slightly larger Co-Co(1) coordination numbers at all the intermediate steps. At the same time, the $N_{Co\text{-}Mo}$ numbers are smaller when the catalyst is sulfided first at atmospheric pressure.

The EXAFS fit parameters of Co(2.25)Mo(7)P(1.18)/Al_2O_3 in Table 6.13 show a Co-S contribution at a distance of 2.22 Å , two Co-Co contributions at about 2.56 Å and one at 3.48 Å at intermediate temperatures. These distances are relatively close to those of bulk Co_9S_8. Upon treatment at 573 K, the Co-Mo contribution is observed in the spectrum and the coordination number ($N_{Co\text{-}Mo}$) does not change appreciably upon subsequent sulfidation at 673 K and 4 MPa.

6.3.4 TEM

Figure 6.10 shows representative TEM micrographs of the Co(2.25)Mo(7)/Al_2O_3 NTA:Mo=1.2 catalyst sulfided at 673 K and different pressures. The dispersion and morphology parameters are summarized in Table 6.14. The stacking degree is higher for the catalyst sulfided at 673 K and 4 MPa. The slabs are also better defined (increased visibility of *d*-spacing) indicating increased crystallinity after the high-pressure treatment. Only a small increase of the stacking degree is observed when raising the sulfidation pressure from 0.1 to 4 MPa.

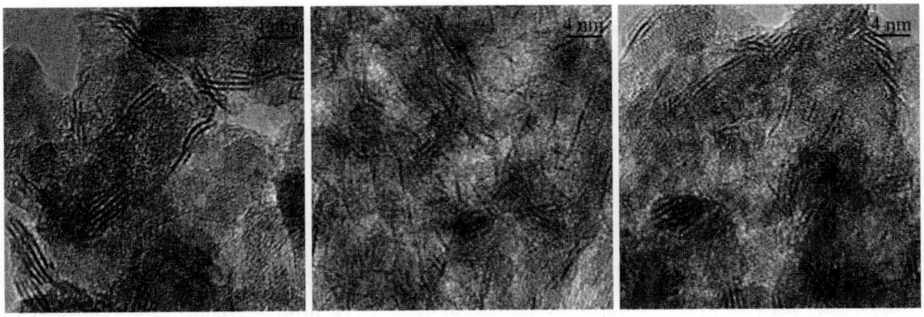

Figure 6.10 *TEM images of Co(2.25)Mo(7)/Al_2O_3 NTA:Mo=1.2 catalyst after various sulfidation treatments. Left: (S, 4 MPa, 673 K); middle: (S, 0.1 MPa, 673 K); right: (S, 0.1 MPa, 673 K + S, 4 MPa, 673 K).*

Table 6.14 *Slab length and stacking degree of Co(2.25)Mo(7)/Al_2O_3 NTA:Mo=1.2 catalyst after different sulfidation treatments*

Treatment	*Average slab length (nm)*	*Average stacking*
S, 4 MPa, 673 K	3.2	1.8
S, 0.1 MPa, 673 K	3.1	1.5
S, 0.1 MPa, 673 K + S, 4 MPa, 673 K	3.1	1.6

The mean slab length and the stacking degree of Co(2.25)Mo(7)/Al_2O_3 NTA:Co=1 sulfided at 673 K and different pressures (Fig. 6.11) are given in Table 6.15. Again, the slabs are better defined when the catalyst is activated at 4 MPa. The slab lengths are very similar for al the three sulfidation treatments.

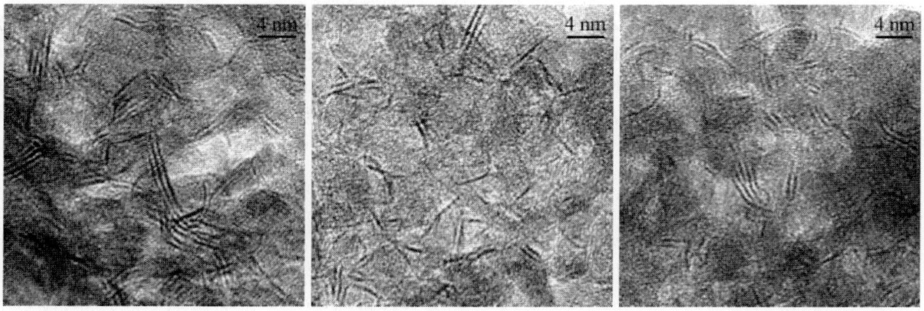

Figure 6.11 *TEM images of Co(2.25)Mo(7)/Al_2O_3 NTA:Co=1 catalyst after various sulfidation treatments. Left: (S, 4 MPa, 673 K); middle: (S, 0.1 MPa, 673 K); right: (S, 0.1 MPa, 673 K + S, 4 MPa, 673 K).*

Table 6.15 *Slab length and stacking degree of Co(2.25)Mo(7)/Al_2O_3 NTA:Co=1 catalyst after different sulfidation treatments*

Treatment	*Average slab length (nm)*	*Average stacking*
S, 4 MPa, 673 K	3.0	1.6
S, 0.1 MPa, 673 K	2.9	1.5
S, 0.1 MPa, 673 K + S, 4 MPa, 673 K	2.9	1.5

Figure 6.12 shows a representative TEM micrograph of the Co(2.25)Mo(7)P(1.18)/Al_2O_3 catalyst sulfided at 673 K and 4 MPa. The results of the statistical analysis of TEM micrographs of this catalyst are summarized in Table 6.16. Clearly, the slabs in this sample are slightly larger and less well-defined compared to the similar calcined catalyst prepared without phosphorus (see section 3.3.4).

Table 6.16 *Slab length and stacking degree of Co(2.25)Mo(7)P(1.18)/Al_2O_3 catalyst after sulfidation treatment*

Treatment	*Average slab length (nm)*	*Average stacking*
S, 4 MPa, 673 K	3.2	1.4

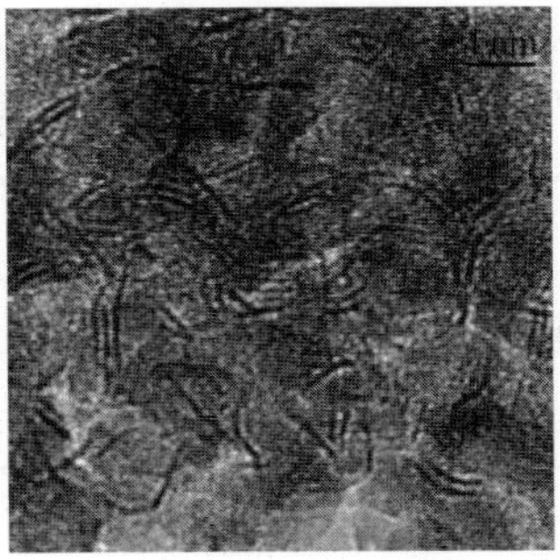

Figure 6.12 *TEM image of $Co(2.25)Mo(7)P(1.18)/Al_2O_3$ catalyst after sulfidation at 673 K and 4 MPa.*

6.3.5 Dibenzothiophene HDS

The DBT HDS activities of the $Co(2.25)Mo(7)/Al_2O_3$ NTA:Mo=1.2 catalyst activated at 673 K and different pressures are collected in Table 6.17. The apparent activation energies (E_{act}) and the pre-exponential factors (ν_{pre}) were determined in the 533-573 K temperature interval. The sulfidation pressure has a large influence on the DBT HDS activity. The activity of the catalyst sulfided at 4 MPa is twice as high as that of the catalysts sulfided at atmospheric pressure. After increasing the sulfidation pressure from 0.1 to 4 MPa the activity of the catalyst also increased substantially. Higher pre-exponential factors are obtained when sulfiding the catalysts under high-pressure conditions.

Table 6.17 *DBT HDS activities, apparent activation energies and pre-exponential factors of $Co(2.25)Mo(7)/Al_2O_3$ NTA:Mo=1.2 catalyst after different sulfidation treatments*

Treatment	k_{HDS} *(mol DBT·Kg $_{Mo}^{-1}$·s^{-1}) ·10^{-2}*	E_{act} *(kJ·mol^{-1})*	ν_{pre} *(mol·kg^{-1}·s^{-1})*
S, 4 MPa, 673 K	3.1	100	$3.7 \cdot 10^7$
S, 0.1 MPa, 673 K	1.5	96	$8.8 \cdot 10^6$
S, 0.1 MPa, 673 K + S, 4 MPa, 673 K	2.5	120	$2.2 \cdot 10^9$

Similar results were obtained after evaluating the catalytic properties of the $Co(2.25)Mo(7)/Al_2O_3$ NTA:Co=1 sample (Table 6.18). The various activities of the catalyst with a lower NTA content were similar to those for the catalyst with NTA:Mo = 1.2.

The determined DBT HDS activities of the $Co(1.3)Mo(7)/Al_2O_3$ NTA:Mo=1.2 sample treated at 673 K and different pressures are shown in Table 6.19. The activity is higher after activation at 0.1 MPa compared to that of the catalyst prepared using less cobalt. The reaction rate constant increases also for the sample subsequently treated at 4 MPa.

Table 6.18 *DBT HDS activities, apparent activation energies and pre-exponential factors of Co(2.25)Mo(7)/Al_2O_3 NTA:Co=1 catalyst after different sulfidation treatments*

Treatment	k_{HDS} (mol DBT·Kg $_{Mo}^{-1}$·s^{-1}) ·10^{-2}	E_{act} (kJ·mol^{-1})	ν_{pre} (mol·kg^{-1}·s^{-1})
S, 4 MPa, 673 K	3.1	102	$5.5 \cdot 10^7$
S, 0.1 MPa, 673 K	1.5	93	$4.9 \cdot 10^6$
S, 0.1 MPa, 673 K + S, 4 MPa, 673 K	2.5	106	$1.2 \cdot 10^8$

Table 6.19 *DBT HDS activities, apparent activation energies and pre-exponential factors of Co(1.3)Mo(7)/Al_2O_3 NTA:Mo=1.2 catalyst after different sulfidation treatments*

Treatment	k_{HDS} (mol DBT·Kg $_{Mo}^{-1}$·s^{-1}) ·10^{-2}	E_{act} (kJ·mol^{-1})	ν_{pre} (mol·kg^{-1}·s^{-1})
S, 0.1 MPa, 673 K	2.3	91	$4.9 \cdot 10^6$
S, 0.1 MPa, 673 K + S, 4 MPa, 673 K	3.5	93	$1.1 \cdot 10^7$

The DBT HDS activities of the Co(1.3)Mo(7)/Al_2O_3 NTA:Co=1 catalyst activated at 673 K and different pressures are shown in Table 6.20. The reaction rate constants are smaller compared to those of the sample prepared using more NTA, but are similar to the rate constants obtained with the Co(2.25)Mo(7)/Al_2O_3 NTA:Co=1 catalyst.

Table 6.20 *DBT HDS activities, apparent activation energies and pre-exponential factors of Co(1.3)Mo(7)/Al_2O_3 NTA:Co=1 catalyst after different sulfidation treatments*

Treatment	k_{HDS} (mol DBT·Kg $_{Mo}^{-1}$·s^{-1}) ·10^{-2}	E_{act} (kJ·mol^{-1})	ν_{pre} (mol·kg^{-1}·s^{-1})
S, 0.1 MPa, 673 K	1.6	73	$7.5 \cdot 10^4$
S, 0.1 MPa, 673 K + S, 4 MPa, 673 K	2.1	73	$9.0 \cdot 10^4$

The results obtained after evaluating the catalytic properties of the Co(2.25)Mo(7)P(1.18)/Al_2O_3 sample are shown in Table 6.21. The determined activity is considerably smaller compared to that of the similar calcined catalyst prepared without phosphorus (see section 3.3.5).

Table 6.21 *DBT HDS activity, apparent activation energy and pre-exponential factor of $Co(2.25)Mo(7)P(1.18)/Al_2O_3$ catalyst after sulfidation at 673 K and 4 MPa*

Treatment	k_{HDS} *(mol DBT·Kg $_{Mo}^{-1}$·s^{-1}) ·10^{-2}*	E_{act} *(kJ·mol^{-1})*	v_{pre} *(mol·kg^{-1}·s^{-1})*
S, 4 MPa, 673 K	2.2	78	$3.1 \cdot 10^5$

6.4 Discussion

After the high-pressure in-situ MES studies on calcined and uncalcined $CoMo/Al_2O_3$ catalysts (Chapters 3 and 4) have shown that sulfidation at elevated pressure leads to predominantly Type II Co-Mo-S formation, we are now also investigating the influence of the pressure on catalysts prepared using additives that decrease the metal-support interactions and should facilitate the Type II phase formation.

The spectra of the fresh samples consist of two quadrupole doublets. The first spectral contribution with an I.S. of 0.20 - 0.30 mm s^{-1} is assigned to Co-oxide species. The second doublet represents a high-spin 2+ phase and is therefore assigned to oxygen containing species, but it can also be related to the formation of the Co-NTA complex [3]. After exposure of the catalysts to the sulfidation mixture at ambient conditions, an increase in the spectral contribution of the high-spin 2+ phase is observed. This can be explained by removal of water associated with the high-spin 2+ species in the fresh catalyst causing an increased interaction of these structures with the support [28]. At intermediate sulfidation steps part of the high-spin 2+ phase can also be assigned to Co-sulfate type species as suggested by previous EXAFS studies [29].

A larger fraction of the Co atoms is present in the high-spin 2+ phase in the fresh catalyst when more NTA is used in the preparation procedure. This indicates that NTA does not preferentially form complexes with the promoter atoms as previously reported [30]. The high-spin 2+ contribution completely disappears from the spectra of the catalysts treated directly at 4 MPa at 373 K. This contribution is still significant at 473 K in the samples activated at 0.1 MPa and points to retarded sulfidation of the Co atoms under these conditions. Although the NTA complex is released much faster at 4 MPa, the sintering of the Co-sulfide species is prevented by the strong Co-Mo interaction induced by the high-pressure treatment [31].

This view is also supported by the increase in the Q.S. values of the Co-sulfide doublet during treatment at 573 K under high-pressure conditions (see Figures 6.3 and 6.6). The better redispersion of the Co-sulfide particles over the edges of the newly formed MoS_2 crystallites, developed at temperatures between 523 K and 573 K [21], points to a stronger Co-Mo interaction in catalysts sulfided at 4 MPa.

No increase of the Q.S. value was observed with the $Co(2.25)Mo(7)P(1.18)/Al_2O_3$ catalyst at 573 K, indicating that the phosphate groups induce a weaker Co-Mo interaction, preventing the redispersion of the Co species over the MoS_2 edges. The decrease in the MoS_2 dispersion and the subsequent decrease of Co dispersion at the edges were previously observed at loadings higher than 0.5 % P [32]. The sulfidation of the high-spin 2+ contribution is identical to that of the similar calcined sample prepared without phosphorus

(Chapter 3), suggesting that the interaction of Co^{2+} cations with the alumina support was not reduced.

After sulfidation at 673 K and 4 MPa, the MES spectra of Co(2.25)Mo(7)/Al_2O_3 NTA:Mo=1.2 and Co(2.25)Mo(7)/Al_2O_3 NTA:Co=1 catalysts (Figures 6.1, G and 6.4, G) were fitted with one quadrupole doublet with relatively low Q.S. values (Q.S. = 0.86 mm s^{-1} and 0.84 mm s^{-1}, respectively). The low Q.S. values of the Co-sulfide doublet point to the partial formation of inactive CoS_x particles. Similar behavior was observed with the uncalcined catalyst having the same metal loadings (Chapter 4), where the weak Co-Mo interaction led to the partial formation of small CoS_x particles having MES parameters that resemble the Co-Mo-S parameters.

Lower Q.S. values (Q.S. ~ 0.80 mm s^{-1} and 0.77 mm s^{-1}, respectively) are obtained when the catalysts are treated also under atmospheric pressure conditions (Figures 6.2, F, G and 6.5, F, G), suggesting the presence of more CoS_x species, or a lower dispersion of the promoter. Although very similar Q.S. values are obtained after activation at 0.1 MPa and after the subsequent increase in the sulfidation pressure, the MES spectra obtained with the samples treated also at 4 MPa show a broader line width (Tables 6.2 and 6.4), indicating a change in the distribution of the local surroundings of Co atoms. A rearrangement of the Co-sulfide particles over the edges of the MoS_2 crystallites, leading to an increased amount of active sites, is proposed to take place upon the subsequent sulfidation at 4 MPa. Variations in the concentration of the promoted sites without significant changes of the active structures nature were recently observed with catalysts prepared with NTA [26].

The Q.S. values obtained with the Co(1.3)Mo(7)/Al_2O_3 NTA:Mo=1.2 catalyst (Table 6.5) are somewhat larger as compared to those of Co(1.3)Mo(7)/Al_2O_3 NTA:Co=1 sample (Table 6.6), pointing to an improved dispersion of the Co atoms in the catalyst prepared using larger amount of NTA. The Co(2.25)Mo(7)P(1.18)/Al_2O_3 catalyst shows the smallest Q.S. value after treatment at 673 K and 4 MPa (Q.S. = 0.75 mm s^{-1}), indicating a more significant presence of the CoS_x structures.

The Co K-edge EXAFS results obtained for Co(2.25)Mo(7)/Al_2O_3 NTA:Mo=1.2 and Co(2.25)Mo(7)/Al_2O_3 NTA:Co=1 (Tables 6.11 and 6.12) show that the Co atoms are sulfided already after treatment at 473 K under atmospheric and high-pressure conditions. After treatment at 473 K and 4 MPa, the Co-Co (1) (2.56 Å) coordination distances are somewhat closer to the values previously reported for Co_9S_8 [24, 33] than after atmospheric sulfidation at 0.1 MPa. This indicates the formation of slightly larger Co-sulfide species at intermediate temperatures, as a result of the higher rate of Co sulfidation at 4 MPa. It was previously proposed that very disordered CoS_x species, with coordination distances and numbers different from those of crystalline cobalt sulfide systems, are formed at the intermediate temperatures [6,29]. After sulfidation at 573 K and 673 K of both NTA-containing samples, high Co-S coordination numbers (5.5 – 5.7) are obtained. This is taken as an indication that no large amounts of Co_9S_8 are formed after these treatments. Also in the MES spectra no Co_9S_8-like doublet with its typical small QS value was observed. The CoS_x structures are formed as soon as the catalysts are exposed to the sulfidation mixture and after activation at 573 and 673 K only part of the Co atoms will participate in the Co-Mo-S species.

The partial redispersion of the Co-sulfide particles at the temperature of MoS_2 crystallites formation is confirmed by the slight decrease of the Co-Co(1) and Co-Co(2) coordination numbers at 573 K. This effect is more evident under high-pressure conditions, when also the increase in the Q.S. values of the Co-sulfide doublet was noticed (Figures 6.3

and 6.6). A similar redispersion model has been proposed for NiS_x particles in NiW catalysts [34].

The EXAFS data analysis of Co(2.25)Mo(7)/Al_2O_3 NTA:Mo=1.2 and NTA:Co=1 catalysts sulfided at 573 K and 673 K reveals the presence of a Co-Mo contribution at 2.76 Å. This value is somewhat smaller than those previously found for samples prepared without NTA (Chapters 3 and 4). The presence of the Co-Mo contribution at 573 K is in line with the conclusion that Co-Mo-S structures are formed at this temperature. The higher Co-Mo coordination numbers obtained with catalysts activated at 4 MPa point in the direction of smaller Co-sulfide species formation (more Co-Mo-S structures) upon increasing the sulfidation pressure as also indicated by the MES spectra.

The Co(2.25)Mo(7)P(1.18)/Al_2O_3 catalyst was sulfided directly under high-pressure conditions and shows the presence of two Co-Co contributions at 2.56 Å and 3.48 Å (Table 6.13). These values are quite close to the corresponding distances in Co_9S_8. This confirms the formation of larger Co-sulfide species in this sample. However, the high Co-S coordination numbers obtained at all temperatures and the MES spectra imply that no formation of bulky Co_9S_8 particles takes place.

The Mo EXAFS fit results for Co(2.25)Mo(7)/Al_2O_3 NTA:Mo=1.2 and Co(2.25)Mo(7)/Al_2O_3 NTA:Co=1 sulfided under both atmospheric and high pressure conditions (Tables 6.8 and 6.9) reveal the presence of relatively large Mo-S coordination numbers ($N_{Mo\text{-}S}$ = 5.3 – 5.8), in combination with short Mo-Mo distances (2.83 Å), after treatment at 473 K. Previous EXAFS studies attributed these contributions to a MoS_3-like intermediate species [22,30,35]. After increasing the sulfidation temperature to 573 K and 673 K, the Mo-S and Mo-Mo interatomic distances are in agreement with the structural parameters of MoS_2 crystallites. Higher Mo-S and Mo-Mo coordination numbers are measured after sulfidation of both NTA samples at 673 K and 4 MPa, indicating an accelerated Mo sulfidation under high-pressure conditions. The better saturation by sulfur of the MoS_2 edges and the formation of somewhat larger crystallites upon sulfidation at 4 MPa indicate the presence of the Type II Co-Mo-S species under these conditions. The somewhat smaller Mo-S and Mo-Mo coordination numbers obtained with samples treated also at 0.1 MPa suggest that only a partial formation of Type II Co-Mo-S has occurred under these conditions [9]. The Mo-Co contributions are observed at about 2.75 Å starting with the treatment at 573 K in good agreement with the Co EXAFS fit results.

The fit results of Mo K-edge EXAFS measurements for Co(2.25)Mo(7)P(1.18)/Al_2O_3 catalyst sulfided at high-pressure (Table 6.10) show sulfidation behavior similar to the calcined catalyst prepared without phosphate (Chapter 3). Again, the Type II Co-Mo-S phase formation is proposed to take place upon activation at 673 K and 4 MPa, although a partial formation of the Type I species cannot be ruled out.

The TEM micrographs of Co(2.25)Mo(7)/Al_2O_3 NTA:Mo=1.2 and Co(2.25)Mo(7)/Al_2O_3 NTA:Co=1 catalysts sulfided at 673 K and different pressures (Figures 6.10 and 6.11) show slightly higher average slab lengths and staking degrees for the samples activated directly under high-pressure conditions. The higher sulfidation rate is argued to induce some sintering of MoS_2 slabs [36]. No significant differences in the dispersion and the morphology of MoS_2 slabs are observed for the atmospheric pressure sulfided samples after the subsequent activation at 4 MPa. For the catalysts treated directly at 4 MPa, better-defined slabs are observed – indicating better crystallinity of the MoS_2 phase upon the higher sulfidation pressure. This confirms to some extent the conclusion that the complete transition to Type II Co-Mo-S species takes place only after the 4 MPa treatment.

The TEM micrograph of Co(2.25)Mo(7)P(1.18)/Al_2O_3 catalyst sulfided at 673 K and 4MPa (Figure 6.12, Table 6.16) shows similar stacking degree to that obtained with the calcined Co(2.25)Mo(7)/Al_2O_3 prepared without phosphate. However, the phosphate-support interaction leads to a weaker interaction of Mo with the support and consequently to some loss in the MoS_2 dispersion.

The DBT HDS activities of the Co(2.25)Mo(7)/Al_2O_3 NTA:Mo=1.2 and Co(2.25)Mo(7)/Al_2O_3 NTA:Co=1 (Tables 6.17 and 6.18) show higher reaction rate constants for the samples treated also at 4 MPa compared to the samples that were activated only at atmospheric pressure. The differences derive from the presence of larger fractions of CoS_x species in catalysts sulfided also at 0.1 MPa and from the partial formation of the Type II Co-Mo-S under atmospheric pressure conditions.

The somewhat higher activities obtained with the Co(1.3)Mo(7)/Al_2O_3 NTA:Mo=1.2 sample (Table 6.19) can be explained by the absence of the CoS_x structures, which can sterically block the active sites, in the lower-loading catalyst. The reaction rate constants are smaller for the Co(1.3)Mo(7)/Al_2O_3 NTA:Co=1 sample (Table 6.20), because of the lower dispersion of the Co atoms in the catalyst prepared using smaller amount of NTA, as indicated by the MES results.

The activity of the Co(2.25)Mo(7)P(1.18)/Al_2O_3 catalyst obtained after activation at 673 K and 4MPa (Table 6.21) is considerably smaller compared to that of the similar calcined sample prepared without phosphate (Chapter 3), because of the significant presence of the CoS_x structures in this catalyst.

The apparent activation energies are slightly higher for the catalysts activated also under high-pressure conditions, behavior understood in terms of increasing surface coverage by dibenzothiophene or H_2S [37]. A strong interaction of the active species with the adsorbed complex can also explain the high pre-exponential factors obtained with these samples. The large changes in the pre-exponential factors could also be related to differences in the number of active sites formed [38].

In an attempt to better explain the large differences in activity observed for samples with similar Q.S. values of the Co-sulfide doublet, an alternative fit with a distribution in Q.S. values was performed, accounting for the presence of active species involving variable number of Co atoms. MossWinn 3.0i [27] was used to analyze the MES spectra obtained with the Co(2.25)Mo(7)/Al_2O_3 NTA:Mo=1.2 catalyst after treatments at 673 K and different pressures, to obtain the distribution of the Q.S. values (Figure 6.13).

It can be observed that after atmospheric pressure activation, the fraction of Co-sulfide species with lower Q.S. values (corresponding to the inactive CoS_x structures) is higher. After increasing the sulfidation pressure, a larger amount of Co-species with MES parameters characteristic of Co-Mo-S phases (QS = 0.9 – 1.3 mm s^{-1}) [2, 25] are formed. The smallest fraction of CoS_x species is obtained when the sample is treated directly at 4 MPa.

Similar trends are found also with the Co(2.25)Mo(7)/Al_2O_3 NTA:Co=1 catalyst after activation at 673 K (Figure 6.14). The fractional behavior can be again qualitatively explained by the formation of larger Co-sulfide species upon sulfidation at 0.1 MPa. No significant changes in the distribution of the Q.S. values were observed with the low promoter-loading catalysts.

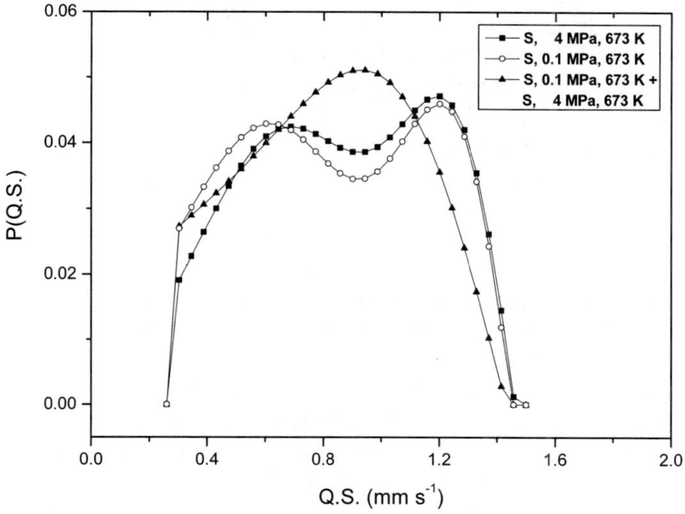

Figure 6.13 *Distribution of Q.S. values of the Co-sulfide doublet obtained with Co(2.25)Mo(7)/Al$_2$O$_3$ NTA:Mo=1.2 catalyst at 673 K and 0.1 MPa (○), 4 MPa (■) and at 0.1 MPa + 4 MPa (▲).*

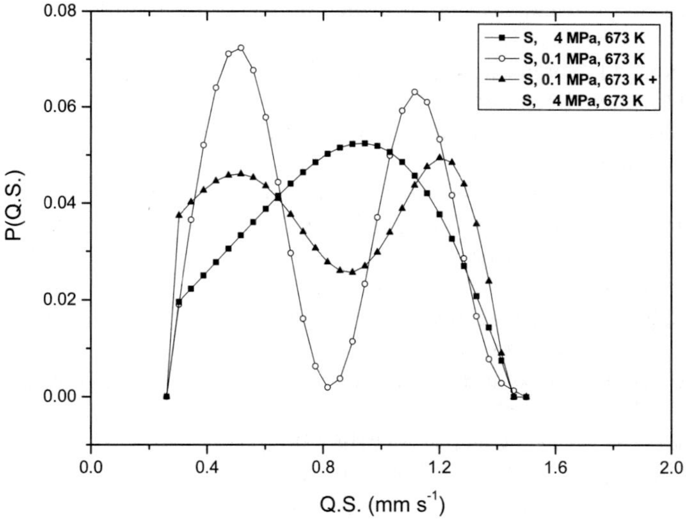

Figure 6.14 *Distribution of Q.S. values of the Co-sulfide doublet obtained with Co(2.25)Mo(7)/Al$_2$O$_3$ NTA:Co=1 catalyst at 673 K and 0.1 MPa (○), 4 MPa (■) and at 0.1 MPa + 4 MPa (▲).*

6.5 Conclusions

The sulfidation of NTA and phosphorus containing catalysts supported on alumina, under conditions relevant to industrial practice, was studied by MES, EXAFS, TEM and DBT HDS.

The presence of additives that decrease the metal-support interactions leads to a significant formation of small inactive CoS_x particles (highly dispersed Co_9S_8 type structures) in the high promoter-loading catalyst, having MES parameters that resemble the Co-Mo-S parameters. By fitting the MES spectra with a distribution of the Q.S. values, accounting for the presence of Co-sulfide species involving variable number of Co atoms, the formation of more CoS_x species in samples also treated at 0.1 MPa is observed. Only the Co-Mo-S species are formed in the lower-loading catalysts, but an improved dispersion of the Co atoms in the sample prepared using larger amount of NTA was noticed. A more significant presence of the CoS_x structures was observed with the P-containing catalyst, where also a weaker interaction of Mo with the support led to some loss in the MoS_2 dispersion as compared to the similar calcined sample prepared without phosphate.

The Mo EXAFS and the TEM data suggest that a complete formation of Type II Co-Mo-S species takes place only after the high-pressure treatment. Higher activities are obtained with catalysts treated also at 4 MPa compared to the samples that are activated only at atmospheric pressure.

References

1. K. Inamura, K. Uchikawa, S. Matsuda, Y. Akai, Appl. Surf. Sci. 121/122 (1997) 468.
2. R. Candia, O. Sorensen, J. Villadsen, N.-Y. Topsøe, B.S. Clausen, H. Topsøe, Bull. Soc. Chim. Belg. 93 (1984) 763.
3. J.A.R. van Veen, E. Gerkema, A.M. van der Kraan, A. Knoester, J. Chem. Soc. Chem. Comm. (1987) 1684.
4. S.M.A.M. Bouwens, F.B.M. van Zon, M.P. van Dijk, A.M. van der Kraan, V.H.J. de Beer, J.A.R. van Veen, D.C. Koningsberger, J. Catal. 146 (1994) 375.
5. J.A.R. van Veen, E. Gerkema, A.M. van der Kraan, P.A.J.M. Hendriks and H. Beens, J. Catal. 133 (1982) 112.
6. E.J.M. Hensen, V.H.J. de Beer, J.A.R. van Veen, R.A. van Santen, Catal. Lett. 84 (2002) 59.
7. A.M. de Jong, V.H.J. de Beer, J.A.R. van Veen, J.W. Niemantsverdriet, J. Phys. Chem. 100 (1996) 17722.
8. L. Medici, R. Prins, J. Catal. 163 (1996) 38.
9. L. Coulier, V.H.J. de Beer, J.A.R. van Veen, J.W. Niemantsverdriet, J. Catal. 197 (2001) 26.
10. J.A.R. van Veen, H.A. Colijn, P.A.J.M. Hendriks, A.J. van Welsenes, Fuel. Proc. Technol. 35, (1993) 137.
11. R. Iwamoto, J. Grimblot, Adv. Catalysis 44 (1999) 417.
12. D. Nicosia, R. Prins, J. Catal. 231 (2005) 259.
13. M. Sun, D. Nicosia, R. Prins, Catal. Today 86 (2003) 173.
14. A. Morales, M.M. Ramirez de Agudelo, Appl. Catal. 23 (1986) 23.

15. P. Atanasova, T. Halachev, J. Uchytil, M. Kraus, Appl. Catal. 38 (1988) 235.
16. P. J. Mangnus, J. A. R. Van Veen, S. Eijsbouts, V. H. J. De Beer, J. A. Moulijn, Appl. Catal. 61 (1990) 99.
17. S. Eijsbouts, J. N. M. Van Gestel, J. A. R. Van Veen, V. H. J. De Beer, R. Prins, J. Catal. 131 (1991) 412.
18. D. Nicosia, R. Prins, J. Catal. 229 (2005) 424.
19. A.I. Dugulan, M.W.J. Crajé, G.J. Kearley, J. Catal. 222 (2004) 281.
20. T.G. Parham, R.P. Merrill, J. Catal. 85 (1984) 295.
21. M. de Boer, A.J. van Dillen, D.C. Koningsberger, J.W. Geus, Jpn. J. Appl. Phys. 32 (1992) 460.
22. R. Cattaneo, T. Weber, T. Shido, R. Prins, J. Catal. 191 (2000) 225.
23. D. Nicosia, R. Prins, J. Catal. 231 (2005) 259.
24. S.M.A.M. Bouwens, J.A.R. van Veen, D.C. Koningsberger, V.H.J. de Beer, R. Prins, J. Phys. Chem. 95 (1991) 123.
25. M.W.J. Crajé, Ph.D. Thesis, Delft University of Technology, Delft, 1992, ISBN 90-73861-08-X.
26. M.A. Lélias, J. van Gestel, F. Maugé, J.A.R. van Veen, Effect of NTA addition on the formation, structure and activity of the active phase of cobalt–molybdenum sulfide hydrotreating catalysts, Catal. Today (2007).
27. Z. Klencsár, Nucl. Instr. Meth. B 129 (1997) 527.
28. M.W.J. Crajé, E. Gerkema, V.H.J. de Beer, A.M. van der Kraan, Hyp. Int. 57 (1990) 1795.
29. M.W.J. Crajé, S.P.A. Louwers, V.H.J. de Beer, R. Prins, A.M. van der Kraan, J. Phys. Chem. 96 (1992) 5445.
30. R. Cattaneo, F. Rota, R. Prins, J. Catal. 199 (2001) 318.
31. N. Koizumi, M. Yamazaki, S. Hatanaka, M. Yamada, Catal. Today 39 (1997) 33.
32. C. Kwak, M.Y. Kim, K. Choi, S.H. Moon, Appl. Catal. A 185 (1999) 19.
33. W. Niemann, B.S. Clausen, H. Topsøe, Catal. Lett. 9 (1990) 355.
34. E.J.M. Hensen, Y. van der Meer, J.A.R. van Veen and J.W. Niemantsverdriet, Appl. Catal. A, 322 (2007) 16.
35. S.P. Cramer, K.S. Liang, A.J. Jacobson, C.H. Chang, R.R. Chianelli, Inorg. Chem. 23 (1984) 215.
36. P.J. Kooyman, J.G. Buglass, H.R. Reinhoudt, A.D. van Langeveld, E.J.M. Hensen, H.W. Zandbergen and J.A.R. van Veen, J. Phys. Chem. B 106 (2002) 11795.
37. E. J. M. Hensen, M. J. Vissenberg, V. H. J. de Beer, J. A. R. van Veen, R. A. van Santen, J. Catal. 163 (1996) 429.
38. E. J. M. Hensen, Ph.D. Thesis, Eindhoven University of Technology, Eindhoven, 2000, ISBN 90-386-2871-4.

Chapter 7

Formulation of structure-activity relations for Mo-based hydrodesulfurization catalysts

Abstract

Structure-activity relations are derived for the series of alumina-supported CoMo catalysts presented in chapters 3, 4 and 6, with the aim to explain the effect of high-pressure sulfidation on the active phase composition and the HDS activity. The spectroscopic studies of the Co promoter phase do not evidence substantial differences to explain the activity trends. Instead, the existence of a continuum of Co-Mo-S structures ranging from a less active Type I phase having a strong interaction with the support up to a fully sulfided and possibly well-crystallized Type II Co-Mo-S phase with a weak support interaction is proposed for sulfided CoMo catalysts. The use of NTA and high-pressure sulfidation leads to the highest Type II character of the active Co-Mo-S phase. A good correlation between the gas-phase thiophene and DBT HDS was observed for most catalysts. An exception is the high DBT HDS activity for a calcined CoMo catalyst sulfided at high pressure. Complementary data support the view that a low stacking degree is the reason for the good catalytic performance in DBT HDS. In contrast, this morphological parameter does not appear to affect the thiophene HDS activity substantially. A low stacking degree of the MoS_2 phase and full sulfidation of Mo are proposed for highly active CoMo sulfide catalysts.

7.1 Introduction

In chapters 3-6 the influence of the sulfidation pressure on the structure and activity has been studied for a set of alumina- and carbon-supported CoMo catalysts. ^{57}Co Mössbauer emission spectroscopy has been used to determine the active phase structure in the sulfided catalysts. Complementary EXAFS measurements at the Co and Mo K-edges show that an increase of the sulfidation pressure results in increased sulfidation degree of the catalysts. From the evolution of the Mössbauer spectra as a function of the sulfidation temperature and pressure additional insight into the sulfidation process has been obtained.

The various spectroscopic data probing the environment around Co (MES, EXAFS) do not reveal significant differences in the final state of the promoter phase. Therefore, the nature of the Co promoter phase does not appear to be the crucial parameter explaining the strong activity differences in DBT HDS noted for the sulfided CoMo catalysts. A natural alternative starting point is then that differences in the degree of Mo sulfidation play a role. Indeed, it is very well known that the degree of Mo sulfidation and, related to this, the type I/II behavior of the active Co-Mo-S structures strongly influences HDS activity. A straigthforward explanation for the increased DBT HDS activities of such better sulfided catalysts has been that sulfidation at 4 MPa transforms a Type I Co-Mo-S phase into a Type II one. Such a trend has also been noted earlier for a commercial alumina-supported CoMo catalyst sulfided at 1 MPa [1]. In this chapter, an attempt is made to consolidate this view by a thorough comparison of the data for the various catalysts. Also, for comparative purposes, some thiophene HDS tests were conducted, and some further catalysts were prepared and tested.

7.2 Experimental

One NTA-containing CoMo/Al_2O_3 catalyst with a lower Co/Mo ratio was prepared by pore-volume impregnation using γ-Al2O3 (Ketjen 001-1.5E, surface area 271 m^2 g^{-1}, pore volume 0.7 ml g^{-1}, particle size 0.5-0.85 mm) as support material. Ammoniacal solution containing ammonium heptamolybdate $(NH_4)_6Mo_7O_{24}.4H_2O$ (Merck, min. 99.9%), cobalt nitrate $Co(NO_3)_2.6H_2O$ (Merck p.a.) and NTA were used in an one-step impregnation procedure as described by Van Veen et al. [2]. After impregnation the catalyst was dried. The catalyst is denoted as Co(0.6)Mo(7)/Al_2O_3(NTA:Mo=1.2). A further set of NiMo catalysts was prepared by pore-volume impregnation of a number of supports with an ammoniacal solution containing the same ingredients as for the Co(0.6)Mo(7)/Al_2O_3(NTA:Mo=1.2) catalyst. As supports were used γ-Al_2O_3 (Ketjen 001-1.5E), SiO_2 (Shell, surface area 210 m^2 g^{-1}, pore volume 1.25 ml g^{-1}) and a set of amorphous silica-aluminas. These supports were prepared by deposition-precipitation of aluminium to silica with final SiO_2/Al_2O_3 between 4 and 19. The final Mo loading in these catalysts was 8 wt.% with a Ni/Mo and NTA/Mo ratios of 0.3 and 1.2, respectively.

Atmospheric-pressure gas-phase thiophene HDS was carried out in a single-pass stainless-steel reactor with an internal diameter of 4 mm. The catalyst was diluted with an appropriate amount of SiC to attain plug-flow conditions. Prior to reaction, the catalysts were sulfided in a mixture of 10 vol.% H_2S in H_2 at a total flow rate of 60 ml min^{-1} at a total pressure of 0.1 or 4 MPa. The NTA-containing catalyst precursors were sulfided at a rate of 2 K min^{-1}. After sulfidation, the reactor was closed with inline valves and transferred to the thiophene HDS reaction setup. The reaction was started at a temperature of 673 K and after reaching steady-state activity after 13 h at 673 K in a mixture of 4 vol.% thiophene in H_2, the temperature was lowered to 623 K. The final thiophene conversion was below 15%. From the steady-state conversion at 623 K the intrinsic reaction rate expressed per mol of Mo was calculated.

7.3 Results and discussion

Alumina-supported CoMo catalysts have been prepared at Co/Mo ratios of 0.3 and 0.52 and occasionally at the very low ratio of 0.01. It is generally believed that Co/Mo ratios in excess of 0.3 lead to partial segregation of a separate Co-sulfide phase, which does not contribute substantially to the catalytic activity [3,4]. For this reason, the discussion starts with a comparison of various catalysts prepared at a Co/Mo ratio of 0.3 despite the somewhat limited amount of characterization data for this set. In a next step, a rationalization of the activities of catalysts with a higher Co/Mo ratio is presented.

After sulfidation at 0.1 MPa, Co(1.3)Mo(7)/Al_2O_3(NTA:Mo=1.2) is about twice as active in DBT HDS as Co(1.3)Mo(7)/Al_2O_3-383 ($2.3.10^{-2}$ mol·kg $^{-1}$·s^{-1} *vs.* $0.9.10^{-2}$ mol·kg $^{-1}$·s^{-1}). Lowering the NTA concentration in Co(1.3)Mo(7)/Al_2O_3(NTA:Co=1) leads to a somewhat lower HDS activity ($1.6.10^{-2}$ mol·kg $^{-1}$·s^{-1}). The large activity differences between these three catalysts are accompanied by virtually identical Q.S. values. The value of the most active catalyst (Q.S. = 1.08 mm s^{-1}) is only marginally higher than that of the lesser active ones (Q.S. = 1.03 and 1.04 mm s^{-1} for the NTA and non-NTA catalyst, respectively). These values fall well in the interval of Q.S. values, which have been used as an indicator to the presence of Co-Mo-S-type phases [5,6]. The large difference between the NTA-containing catalysts and the dried catalyst should undoubtedly be ascribed to the Co-Mo-S phase being, at least partly, present as Type II [4] in the former. Earlier work had already indicated that ^{57}Co MES cannot distinguish between Type I and II Co-Mo-S phases [4,7,8] and the present dataset supports this view.

The active phase in Co(1.3)Mo(7)/Al_2O_3(NTA:Mo=1.2) catalyst sulfided at 0.1 MPa is regarded as the archetype of the Co-Mo-S phase II because the metal sulfide phase appears to be fully sulfided now [2]. By resolving the local structure around Mo and Co with EXAFS one finds a trend of increased Mo-S and Co-S coordination numbers going from Type I to Type II Co-Mo-S [9]. These changes are in agreement with a higher degree of sulfidation of the active components and specifically the removal of some remaining covalent bonds with the subjacent alumina support by sulfidation has been argued to be crucial in the transition of a Type I active phase into the more active Type II one. Indeed, the Mo-S coordination number within the present set of catalysts increases from 5.1 (dried catalyst) to 5.7 (NTA:Co=1) and 5.9 (NTA:Mo=1.2). It should be noted that these coordination numbers have been obtained for the Co(2.25)Mo(7)/Al_2O_3 catalysts. Although such differences in the sulfidation degree also present themselves in the Co-S coordination numbers in typical CoMo catalysts with close-to-optimal Co/Mo ratios of around 0.3, for the Co(2.25)Mo(7)/Al_2O_3 set the differences in this parameter are very small which should be due to the considerably higher Co loading.

It is now essential to include the results of the same Co(1.3)Mo(7)/Al_2O_3 catalysts after sulfidation at 4 MPa. The Q.S. values do not change compared to those after sulfidation at 0.1 MPa and this strongly suggests that the particle size or ordering of the promoter sulfide phase has not changed substantially. Also, there are no significant differences in the particle size of the MoS_2 slabs as determined from by electron microscopy, albeit that there is an increase of the stacking degree to a value of about 1.8. Nevertheless, the DBT HDS activity of the various catalysts is strongly improved. For individual catalysts in preceding chapters, this effect has been explained in terms of phase I transforming into phase II after high-pressure sulfidation and the higher stacking degree tallies with such an explanation. Most peculiar however it is found that the activity increase is highest for the NTA:Mo=1.2 catalyst which was argued to contain already an optimal Type II Co-Mo-S phase.

An alternative view on all this starts with the hypothesis that sulfidation of the NTA:Mo=1.2 catalyst at 0.1 MPa does not lead to a fully sulfided Type II Co-Mo-S phase but instead leaves the active phase somewhere between Type I and II. Sulfidation at 4 MPa apparently leads to an even better sulfided catalyst with a much higher intrinsic activity. Evidence for the higher sulfidation degree of the MoS_2 phase is found in the increase of the Mo-S coordination number of Co(2.25)Mo(7)/Al_2O_3(NTA:Mo=1.2) 5.8 (0.1 MPa) to 6.1 (4 MPa). This line of reasoning suggests that the non-NTA and NTA:Co=1 catalysts should contain an active phase of a substantial Type I character, even after sulfidation at 4 MPa.

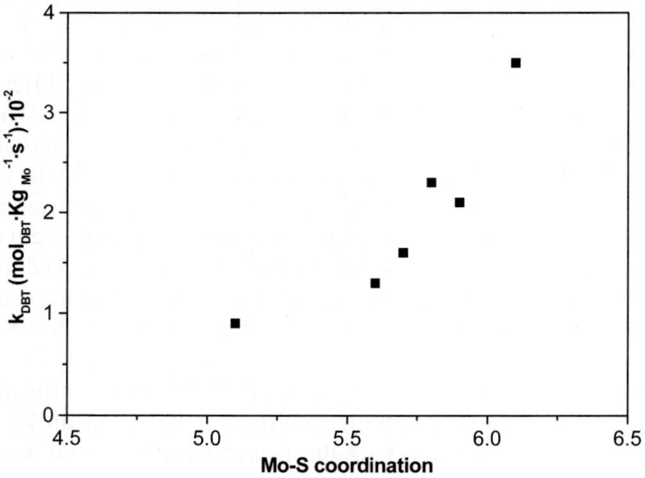

Figure 7.1 *Relation between DBT HDS activity and Mo-S coordination number for the Co(1.3)Mo(7)/Al_2O_3 catalysts (dried, NTA:Mo=1.2 and NTA:Co=1) after sulfidation at 0.1 and 4 MPa.*

Fig. 7.1 shows that there is a strong correlation between the DBT HDS activity and Mo-S coordination numbers for this set of catalysts. It is reasonable to state that this Mo-S coordination number correlates to the sulfidation degree of the MoS_2 phase and in turn one may see this as an indication of the approach of the active Co-Mo-S phase to a true Type II one. A first consequence is that the NTA:Mo=1.2 catalyst sulfided at 0.1 MPa should not be regarded as the Type II end member of a range of Co-Mo-S structures spanning the Type I/II range. One has to accept furthermore that there may be a gradual change between the pure Type I and II end members of such a series as already suggested earlier for alumina-supported MoS_2 catalysts [10]. The Mo-S coordination number of the NTA:Mo=1.2 catalyst sulfided at 4 MPa of 6.1 strongly suggests that sulfidation of the MoS_2 phase is complete and it is now useful to regard its active Co-Mo-S phase as a pure Type II one. It then follows that sulfidation of Co(1.3)Mo(7)/Al_2O_3-383 at 4 MPa still leaves the active phase largely in a Type I state. This supposition is fortified by noting that the Mo-S coordination number (5.6) after sulfidation at 4 MPa is still below that of the atmospherically sulfided NTA-containing catalysts. Thus, it is much more difficult to fully sulfide the molybdenum oxide phase in this dried catalyst as is well-known for calcined catalysts. This does not appear to be unreasonable since the Mo(7)/Al_2O_3 precursor to this dried catalyst was calcined.

The NTA:Co=1 catalyst takes an intermediate position and judging from the Mo-S coordination numbers the NTA can in this case at least partially induce better Mo sulfidation. Although it is generally accepted that NTA has a preference for the promoter ions in impregnation solutions containing Ni^{2+} ions and molybdate anions [11] – the complexation constants of Ni-NTA and Co-NTA are similar [12] – the ultimate fate of the chelating agent upon drying is unclear. The data suggest that even when there is just enough NTA to complex the Co^{2+} ions in solution an interaction of NTA with the alumina support or Mo phase develops upon drying. To exclude that the DBT HDS data are a testing artifact, the NTA-containing catalysts were also tested in atmospheric-pressure thiophene HDS after sulfidation at 4 MPa. The large difference in DBT HDS is reproduced in thiophene HDS with intrinsic reaction rates of 12 and 27 h^{-1} for the low and high NTA loading, respectively.

Summarizing, the active phase in the three Co(1.3)Mo(7)/Al_2O_3 catalysts after 0.1 MPa or 4 MPa sulfidation consists of Co-Mo-S structures with a well-dispersed Co sulfide promoter phase. The large activity variations within this set can be linked to the degree of Mo sulfidation as derived from the Mo-S coordination numbers. Sulfidation at 4 MPa leads to a higher sulfidation degree and to substantially higher DBT HDS activities, even for a CoMo-NTA catalyst with sufficient NTA (NTA:Mo=1.2) sulfided at 0.1 MPa in which a Type II Co-Mo-S should reign supreme in our conventional thinking. However, the present EXAFS data suggest that MoS_2 sulfidation is only then complete when sufficient NTA in combination with high-pressure sulfidation is applied and tentatively this defines a pure Type II phase end member. Lower amounts of NTA limit the approach of the Co-Mo-S phase to a Type II one, because of some remaining Mo-support interaction. An extreme case is the calcined catalyst in which the strong Mo-support interaction limits the transition of the active metals to a Type II phase. A general conclusion is then that once an interaction of Mo with the alumina support is present it is very difficult to fully sulfide molybdenum oxide even at high pressure. In spite of the known difficulty in accurately determining coordination numbers from X-ray absorption spectra, the Mo-S coordination number might thus prove a useful probe to the Type I/II character of the Co-Mo-S phase, the more since it is not easy to see what other technique might be able to probe these minute differences in the Mo sulfidation degree.

Typically, full sulfidation of NTA:Mo=1.2 catalysts has been derived from the absence of some oxidic or oxysulfidic Mo states which are typically encountered for Type I Co-Mo-S and likely represent remaining Mo-O-Al linkages [3]. Thus, remaining linkages should be so few that they are not amenable to experimental observation anymore when sulfidation has proceeded sufficiently. A somewhat different view on this would be that MoS_2 formation is complete after sulfidation at 0.1 MPa and that further sulfidation at 4 MPa leads to a more crystalline Co-Mo-S (MoS_2) phase. In support of the higher crystallinity are the slightly higher Mo-Mo coordination numbers after high-pressure sulfidation at nearly constant MoS_2 dispersion and the somewhat better definition of the MoS_2 slabs in electron microscopy imaging. Moreover, this would offer an attractive explanation for the little variation of the Q.S. values with sulfidation pressure while at the same time the Co-Mo and Mo-Co coordination numbers increase somewhat after sulfidation at 4 MPa. Kooyman et al. [1] have proposed that a fully sulfided, yet not fully crystallized Co-Mo-S phase is optimal in thiophene HDS, as it was found that sulfidation at 1 MPa leads to a higher HDS activity. It was proposed that a Type I to II transition takes place and this agrees well with the present data. Sulfidation at 873 K instead of 673 K led to a large decrease of the intrinsic thiophene HDS activity and this was suggested to be due to full crystallization of the slabs. This appears to contradict the present view and an alternative not discussed in Ref. [1] is that in fact at the high sulfidation temperature of 873 K some CoS_x sintering has taken place, which impedes catalytic activity. It is comfortable to note that for the discussed set of CoMo catalysts having active Co-Mo-S phases that seem to span at least a large part of the Type I/II forms the Q.S. values of the CoS_x phase do not change appreciably. This suggests that at this close-to-optimal Co/Mo ratio the Co promoter phase remains part of the Co-Mo-S phase and there is no segregation of a separate phase at higher sulfidation pressures. Thus, this set of catalysts spans the range from 0-100% Type II Co-Mo-S phase, where the Type II Co-Mo-S may refer either a fully sulfided active phase or fully sulfided and crystalline active phase.

For the Co(2.25)Mo(7) catalysts the lower Q.S. values indicate a much less optimal dispersion of the cobalt sulfide phase. For the same set of catalysts (dried, NTA:Co=1 and NTA:Mo=1.2) the quadrupole splitting is quite low which suggests that the CoS_x particles have grown beyond a size that is considered to be characteristic for atomically dispersed Co in a Co-Mo-S phase [5]. This can be interpreted in terms of the presence of some CoS_x particles not associated with the Co-Mo-S phase for which the somewhat larger line widths of the quadrupole doublet provide some evidence, but also in terms of larger CoS_x particles on the verge of segregation from the MoS_2 phase. Note that there is no agreement of Q.S. values which constitute the Co-Mo-S phase [5,6]. The fact that the promoter sulfide phase is much less defined in the Co(2.25)Mo(7) samples makes it more difficult to compare the various catalysts. The DBT HDS activity of the dried catalyst is slightly higher than of its Co(1.3) counterpart, especially after sulfidation at 4 MPa. Higher Co/Mo ratios lead in general to somewhat better Mo sulfidation [3]. For the NTA:Co=1 catalyst the activity increases also somewhat which may be attributed to the higher NTA concentration (NTA:Mo=0.52) and consequently a more pronounced Type II character of the active phase. For these catalysts the effect of somewhat less dispersed or unassociated CoS_x particles may be compensated by a more pronounced Type II active phase leading to an overall higher HDS activity. On the other hand, starting from an almost ideal Type II phase in the Co(1.3)Mo(7)/Al_2O_3(NTA:Mo=1.2), the loss of CoS_x dispersion in the catalyst with a higher Co/Mo ratio as evidenced by the much lower Q.S. values (~0.85 vs. 1.08 mm s^{-1}) decreases the DBT HDS activity.

The calcined catalyst Co(2.25)Mo(7)/Al_2O_3-673 sulfided at 0.1 MPa has a DBT HDS activity of $2.0.10^{-2}$ mol·kg $^{-1}$·s^{-1}. The Q.S. value of 0.95 mm s^{-1} indicates that this catalyst contains predominantly a Co-Mo-S phase. The Mo-S coordination number is quite low as may be expected by now for a calcined catalyst and therefore should contain predominantly a Type I Co-Mo-S phase. The DBT HDS activity is however quite high and this can be attributed to the more optimal Co-Mo interaction obtained after calcination than appears possible after drying or even the use of NTA for the Co(2.25)Mo(7) samples. This difference can also be inferred from the higher Q.S. value for the calcined catalyst. Thus, while the edge occupation at Co/Mo=0.3 is close to optimal with an atomically or at least well-dispersed CoS_x phase, it is much more difficult to keep the CoS_x particles in the Co-Mo-S phase at Co/Mo=0.52. The presence of NTA is not of much use to keep Co in the Co-Mo-S phase in this case and one straightforward explanation is that there is just not enough NTA to complex all Co if one indeed accepts that the chelating agent does not remain associated with the Co in the dried catalyst. The activity of the NTA-containing catalysts is higher than that of the dried catalyst because of the higher sulfidation degree of the MoS_2 phase. The most active alumina-supported catalyst then turns out to be the calcined Co(2.25)Mo(7) one sulfided at 4 MPa. Its activity is more than twice that of its atmospherically sulfided counterpart and also substantially higher than of the Co(1.3)Mo(7)/Al_2O_3(NTA:Mo=1.2) catalyst. The Q.S. value does not change going from sulfidation at 0.1 MPa to 4 MPa. The increase in Mo-S coordination number suggests that the active phase has developed a more pronounced phase II character although in terms of the data in Figure 7.1 one would still surmise that the catalyst contains a Type I active phase to some extent.

It is difficult to immediately see what is the cause of the exceptional DBT HDS activity of the calcined catalyst sulfided at 4 MPa. Before speculating on this issue in more detail, the thiophene HDS activities of a subset of the catalysts is presented as it appeared useful to determine the catalytic performance of some catalysts at least in one other test reaction. A first-order approximation is that the gas-phase thiophene HDS activity correlates well with the performance in gas-phase dibenzothiophene HDS [13]. For thiophene HDS experiments, the catalysts were first sulfided at 0.1 or 4 MPa. The intrinsic thiophene HDS activity was then determined under differential conditions at atmospheric pressure. Figure 7.2 compares the activities in thiophene and DBT HDS. Clearly, there is a strong correlation between these two parameters within a certain range except for one catalyst Co(2.25)Mo(7)/Al_2O_3-673 sulfided at 4 MPa.

The good correlation between thiophene and DBT HDS for most of the catalysts is in agreement with the notion that in both reactions most likely C-S hydrogenolysis is rate limiting [14]. Except for catalysts with a low NTA amount (NTA:Co=1) or a low Co/Mo ratio (Co/Mo=0.15), ranking the thiophene HDS activities shows that the NTA-containing catalysts are more active than the non-NTA counterparts. This tallies with the quite large literature database of thiophene HDS activity measurements [2,4]. Gandubert et al. [15] have recently suggested a positive effect of the presence of mixed sites, Co edge atoms next to 'unoccupied' Mo edge atoms, on the catalytic performance in toluene hydrogenation. Since catalysts within a large range of Co/Mo ratios between 0.15 and 0.5 do not show outstanding behavior in Figure 7.2, it is concluded that the presence of such mixed sites does not affect the thiophene HDS activity significantly different than the DBT HDS activity. This result agrees with C-S bond hydrogenolysis being rate-limiting as noted above. Thus, the presence of mixed sites is not expected to have a large effect on the activity in the HDS test reactions. It is also clear that the variations in Q.S. values do not lead to exceptional behavior.

The one catalyst that stands out in the plot in Figure 7.2 remains Co(2.25)Mo(7)/Al_2O_3(4, 673) with a disproportionally high DBT HDS activity. To ascertain that this was no faulty activity measurement, the DBT HDS activity was determined again and a value of 5.0 $mol \cdot kg^{-1} \cdot s^{-1}$ was found.

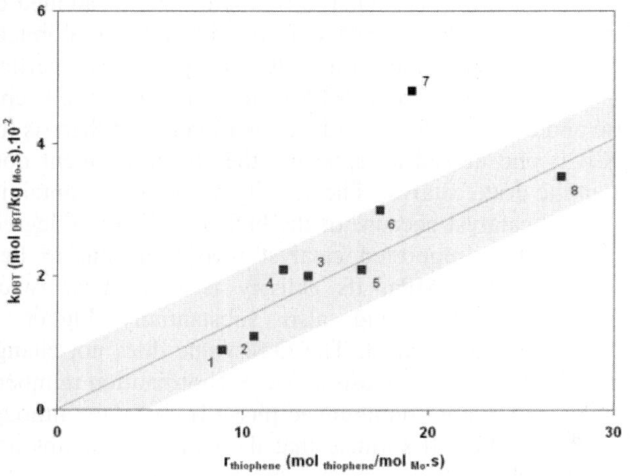

Figure 7.2 *Correlation between first-order rate constant DBT HDS and intrinsic reaction rate for thiophene HDS for (1) Co(1.3)Mo(7)/Al$_2$O$_3$-383(0.1, 673); (2) Co(2.25)Mo(7)/Al$_2$O$_3$-383(0.1, 673); (3) Co(2.25)Mo(7)/Al$_2$O$_3$-673(0.1, 673); (4) Co(1.3)Mo(7)/Al$_2$O$_3$(NTA:Co=1, 4, 673); (5) Co(2.25)Mo(7)/Al$_2$O$_3$-383(4, 673); (6) Co(0.6)Mo(7)/Al$_2$O$_3$(NTA:Mo=1.2, 4, 673); (7) Co(2.25)Mo(7)/Al$_2$O$_3$-673(4, 673) and (8) Co(1.3)Mo(7)/Al$_2$O$_3$(NTA:Mo=1.2, 4, 673).*

The one notable difference with the other high-pressure sulfided catalysts is that sulfidation at 4 MPa does not lead to a higher stacking degree in this catalyst. The higher stacking degree of catalysts prepared by NTA has been generally noted [2,4,10] and is regarded as exemplary of its higher sulfidation degree. Co(2.25)Mo(7)/Al$_2$O$_3$-383 has a stacking degree of 1.8 after sulfidation at 4 MPa. At first sight, this is surprising since the Mo(7)/Al$_2$O$_3$ precursor to which Co was added has been calcined and one would expect similar behavior as for Co(2.25)Mo(7)/Al$_2$O$_3$-673. It is however reasonable to assume that the second impregnation step has affected the Mo oxide phase to some extent, thereby leading to increased stacking after high-pressure sulfidation.

These data strongly point to a relatively strong effect of the stacking degree on the DBT HDS activity whilst this morphological parameter does not affect the thiophene HDS activity very much. Earlier work has attempted to correlate the DBT HDS activities to the MoS$_2$ morphology. Daage and Chianelli [16] have found that the top and bottom MoS$_2$ layers of bulk MoS$_2$ (rim sites) are active in hydrogenation and hydrodesulfurization of DBT whereas the layers in between (edge sites) catalyze preferentially hydrodesulfurization. Hensen et al. [17] have found for supported MoS$_2$ that the HDS rate of DBT increases with the stacking degree. Clearly, these model studies do not provide an adequate explanation for the trends observed for Co-promoted MoS$_2$. This is altogether not so surprising because in the CoMo-based catalysts the occupation of the MoS$_2$ edges by Co-sulfide species should be rather high and one is in fact dealing with a different active phase. Instead, it appears that stacking of Co-promoted MoS$_2$ slabs has an adverse effect on the DBT HDS activity, whereas it does not affect the thiophene HDS activity that much. Thus, the Co-Mo-S slab in direct contact with the subjacent support may be suggested to have a higher activity in DBT HDS than the additional layers on top of the first layer.

Typically, such model Type II systems have been prepared by using chelating agents such as NTA in the precursor. The absence of strong Mo-support interactions in this case results in better sulfidation of the active metals. However, the present data suggest that this also has the negative side effect, namely an increased stacking degree. A better approach, at least for DBT HDS, appears to be sulfidation of a calcined CoMo/Al_2O_3 precursor at elevated pressures. The increased sulfidation rate ensures at least partial formation to Type II Co-Mo-S structures compared to atmospheric pressure sulfidation, but the relatively strong interaction between Mo and the alumina support prevents further stacking of the MoS_2 phase. The catalyst morphology of such a catalyst can be viewed as one frozen in transition to a stacked Type II morphology. Although the sulfidation of the Co promoter ions goes through a less-dispersed CoS_x intermediate phase, these CoS_x particles are redispersed over the edges of the MoS_2 slabs as soon as they are formed. This suggests that there is no segregation of the promoter sulfide phase from the nascent Co-Mo-S structures. Such a sulfidation procedure results in increased structural order in the Co-Mo-S phase compared to sulfidation at 0.1 MPa, as follows from the extensive characterization (TEM, EXAFS), but the sulfidation appears to be less complete as compared to NTA-containing catalysts.

A preliminary conclusion of all this is that the use of NTA and high-pressure sulfidation leads to a more pronounced Type II character of the active Co-Mo-S phase. As generally accepted the Type II phase is more active than the Type I one in gas-phase thiophene and DBT HDS. The increased sulfidation degree also leads to increased stacking of the MoS_2 slabs. An exception to this rule appears to be sulfidation at high pressure of a calcined catalyst. Sulfidation is not complete which should be undoubtedly related to the persistence of some support interactions. The active phase appears to be somewhere between Type I and II Co-Mo-S and while its thiophene HDS activity is not unusual in terms of the trends in Figure 7.2, the resulting DBT HDS activity is exceptionally high. From a comparison of the characterization data, it seems that the stacking degree is the one parameter which stands out for this catalyst.

To obtain some more insight into the effect of stacking, a set of NiMo sulfide catalysts was prepared on supports of varying silica/alumina composition and the catalytic activities in DBT and thiophene HDS were determined [18]. The catalysts were prepared with NTA so as to ensure a constant HDS activity, at least in thiophene hydrodesulfurization [2]. These catalysts were tested in the HDS of DBT and thiophene after atmospheric pressure sulfidation. Table 7.1 shows the morphological parameters from electron micrograph analyses and the HDS activities which have been corrected for the MoS_2 dispersion. The activities are expressed relative to that of the NiMo-NTA/Al_2O_3 catalyst. With increasing silica content of the catalyst support an increase in the stacking degree of the MoS_2 phase is noted. This trend can simply be attributed to the decreased Mo-support interaction which is most explicit for the silica-supported catalyst. As expected, the thiophene HDS activities of the NiMo-NTA catalysts do not vary to a large extent. These catalyst should contain predominantly Type II 'Ni-Mo-S' phases, although one would expect by now that the Type II character of the active phase could be further improved by increasing the sulfidation pressure. However, the DBT HDS activities strongly vary amongst the present set of NiMo catalysts.

Table 7.1 *Morphology and HDS activities of a set of alumina- and amorphous-silica-alumina-supported NiMo-NTA catalysts*

Catalyst	$d_{average}$ *(nm)*	$S_{average}$	$r'_{thiophene}$ [a]	k'_{DBT} [b]
NiMo-NTA/Al_2O_3	2.8	1.4	1.0	1.0
NiMo-NTA/ASA(55/45)	2.9	1.37	0.97	0.98
NiMo-NTA/ASA(20/80)	3.2	1.49	1.06	0.76
NiMo-NTA/ASA(15/85)	3.0	1.44	0.99	0.70
NiMo-NTA/ASA(10/90)	2.8	1.49	0.99	0.73
NiMo-NTA/ASA(5/95)	3.0	1.65	1.02	0.69
NiMo-NTA/SiO_2	3.3	2.02	1.08	0.56

[a] intrinsic HDS activity ($mol_{thiophene} \cdot mol_{Mo}^{-1} \cdot h^{-1}$) relative to activity of NiMo-NTA/Al_2O_3
[b] first-order HDS reaction rate constant ($mol_{DBT} \cdot kg_{Mo}^{-1} \cdot s^{-1}$) relative to activity of NiMo-NTA/Al_2O_3

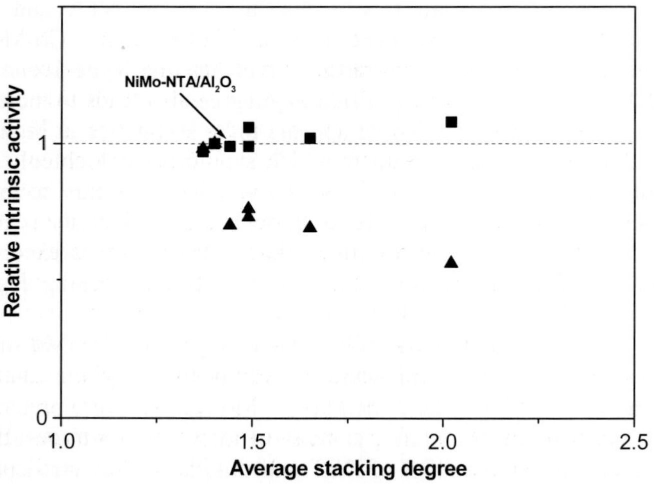

Figure 7.3 *Relative intrinsic DBT HDS (■) and thiophene HDS (▲) activities as a function of the average stacking degree. The reference catalyst is NiMo-NTA/Al_2O_3.*

Figure 7.3 then shows convincingly that the intrinsic activity for DBT HDS decreases strongly with increased stacking degree whereas the influence of stacking on the thiophene HDS activity is very small. One should not *a priori* exclude an (electronic) influence of the support acidity [19], but this does not appear to be of overriding importance here. The strong decrease of the DBT HDS activity with stacking degree suggests that the intrinsic activity of the first layer in direct contact with the support must have a considerably higher DBT HDS activity than subsequent layers on top of the first one. It is reasonable to assume that the first slab in contact with the support has at least some Type I character and this would be then in line with the results for the calcined catalyst.

Finally, it is important to stress that the trends as discussed here are specific for gas-phase DBT and thiophene HDS. That is to say that the trends for other types of hydrotreating reactions such as hydrodenitrogenation or even HDS of the more refractory substituted DBTs may be different. Also the mode of operation of the reaction may influence the noted activity differences. For catalysts activated under atmospheric pressure conditions, it was found that Type II Co(Ni)-Mo-S is much more active in gas-phase HDS reactions than Type I, but slightly less active in trickle-flow DBT HDS [20]. In the presence of a condensed liquid-phase the active sites of Type II catalysts are more affected than those of catalysts with Type I morphology [13]. Different trends were also observed for the HDN and HDS reactions [21]. The presence of more regular MoS_2 crystallites influencing the adsorption and hydrogenation of o-toluidine was the determining factor for the HDN reaction, while a better promoter dispersion played an important role for the HDS of thiophene. As such, the present conclusions should be tested against a wider range of reaction experiments.

7.4 Conclusions

Sulfidation at elevated pressure in a mixture of H_2S/H_2 leads to a better sulfidation of the active Co-Mo-S phase in alumina-supported CoMo catalysts. For calcined catalysts which typically have a Type I active Co-Mo-S phase after sulfidation at atmospheric pressure the HDS activity is increased when the sulfidation pressure is increased. This is due to partial transformation to a more active Co-Mo-S phase with an enhanced Type II character. It remains difficult to fully sulfide the active phase and only when Mo-support interactions are minimized in the precursor by the use of a chelating agent is it possible to prepare a pure Type II Co-Mo-S phase after high-pressure sulfidation. Atmospheric pressure sulfidation of a CoMo-NTA catalyst leads to an active Co-Mo-S phase somewhere between the Type I and II. A reasonable indicator of the Type I/II character of the active phase appears to be the Mo-S coordination number derived from Mo K-edge X-ray absorption spectroscopy. From the large dataset of results, one may abstract the following points regarding the phase I/II behavior of sulfided CoMo catalysts: (i) there is a continuum of active phase Co-Mo-S structures ranging from a less active Type I phase having a strong interaction with the support up to a fully sulfided and possibly well-crystallized Type II Co-Mo-S phase with a weak support interaction; (ii) a pure Type II end member is obtained after high-pressure sulfidation of a CoMo-NTA catalyst, whereas the earlier Type II phase resulting from an atmospherically sulfided CoMo-NTA catalyst should now be regarded as having an active phase with some Type I character.

Concomitant with reaching a full sulfidation degree in NTA catalyst, stacking of the MoS_2 slabs occurs. This stacking does not appear to influence gas-phase thiophene HDS much but impedes activity in gas-phase DBT HDS to some extent. The negative effect of stacking on the DBT HDS activity is reproduced for a set of supported NiMo sulfide catalysts. This could point to some heterogeneity in the distribution of promoter phases on the edges of the slabs in direct contact with the support and the ones on top of that. Thus, the DBT HDS activity of CoMo-NTA catalyst sulfided at high-pressure (pure phase II) with a somewhat stacked MoS_2 morphology may be improved by limiting the stacking of the active phase. The most active catalyst in DBT HDS is a calcined CoMo catalyst and an important explanation seems to be that a large portion of the active phase consists of monolayer slabs. The active phase of this catalyst should be somewhere between a Type I and II Co-Mo-S phase and thus further modifications to increase the sulfidation degree whilst keeping the stacking degree low should be the way to improve the activity. Tentatively, it follows that the requirements of a low stacking degree and full MoS_2 sulfidation are met in carbon-supported CoMo catalysts sulfided at high-pressure which show exceptional activity in DBT HDS.

References

1. P.J. Kooyman, J.G. Buglass, H.R. Reinhoudt, A.D. van Langeveld, E.J.M. Hensen, H.W. Zandbergen and J.A.R. van Veen, J. Phys. Chem. B 106 (2002) 11795.
2. J.A.R. van Veen, E. Gerkema, A.M. van der Kraan and A. Knoester, J. Chem. Soc., Chem. Commun. (1987) 1684.
3. H. Topsøe, B.S. Clausen and F.E. Massoth, Hydrotreating catalysis (Springer, Berlin, 1996).
4. J.A.R. van Veen, E. Gerkema, A.M. van der Kraan, P.A.J.M. Hendriks and H. Beens, J. Catal. 133 (1992) 112.
5. H. Topsøe, R. Candia, N.-Y. Topsøe, B.S. Clausen, Bull. Soc. Chim. Belg. 93 (1984) 783.
6. M.W.J. Crajé, V.H.J. de Beer, J.A.R. van Veen, A.M. van der Kraan, in: M.L. Occelli, R.R. Chianelli (Eds.), Hydrotreating Technology for Pollution Control, Dekker, New York, 1996, p. 95.
7. H. Topsøe, R. Candia, N.-Y. Topsøe and B.S. Clausen, Bull. Soc. Chim. Belg 93 (1984) 783.
8. H. Topsøe and B.S. Clausen, Appl. Catal. 25 (1986) 273.
9. S.M.A.M. Bouwens, F.B.M. van Zon, M.P. van Dijk, A.M. van der Kraan, V.H.J. de Beer, J.A.R. van Veen and D.C. Koningsberger, J. Catal. 146 (1994) 375.
10. E.J.M. Hensen, V.H.J. de Beer, J.A.R. van Veen and R.A. van Santen, Catal. Lett. 84 (2002) 59.
11. L. Medici and R. Prins, J. Catal. 163 (1996) 28.
12. Y. Ohta, T. Shimizu, T. Honma, M. Yamada, Stud. Surf. Sci. Catal. 127 (1999) 161.
13. H.R. Reinhoudt, C.H.M. Boons, A.D. van Langeveld, J.A.R. van Veen, S.T. Sie and J.A. Moulijn, Appl. Catal. A 207 (2001) 25.
14. T. Weber and J.A.R. van Veen, Catal. Today 130 (2008) 170.
15. A.D. Gandubert, E. Krebs, C. Legens, D. Costa, D. Guillaume and P. Raybaud, Catal. Today 130 (2008) 149.
16. M. Daage and R.R. Chianelli, J. Catal. 149 (1994) 414.
17. E.J.M. Hensen, P.J. Kooyman, A.M. van der Kraan, Y. van der Meer, V.H.J. de Beer and R.A. van Santen, J. Catal. 199 (2001) 224.
18. D.G. Poduval, unpublished results.
19. C.E. Hédoire, C. Louis, A. Davidson, M. Breysse, F. Maugé and M. Vrinat, J. Catal. 220 (2003) 433.
20. J.A.R. van Veen, H.A. Colijn, P.A.J.M. Hendriks, A.J. van Welsenes, Fuel. Proc. Technol. 35, (1993) 137.
21. R. Cattaneo, F. Rota, R. Prins, J. Catal. 199 (2001) 318.

Chapter 8

Effect of pressure on the sulfidation behavior of NiW catalysts: a ^{182}W Mössbauer spectroscopy study

Abstract

The sulfidation of Al_2O_3- and ASA- supported NiW catalysts under conditions relevant to industrial practice was studied for the first time by ^{182}W Mössbauer spectroscopy. Only limited number of ^{182}W Mössbauer experiments have been performed previously, mainly due to the need of having specialized equipment for such measurements. ^{182}W MAS can clearly distinguish between WO_3- and WS_2-type phases encountered in the calcined and sulfided forms of W-based hydrotreating catalysts. NiW/Al_2O_3 catalysts are more difficult to sulfide than their Mo-based counterparts and, hence, intermediate stages of sulfidation can be studied as separate phases. At low sulfidation rates, an intermediate WS_3-type phase is identified at temperatures as high as 673 K. The presence of Ni atoms facilitates sulfidation and in this case the formation of an oxysulfidic intermediate is observed. Sulfidation at 673 K and 0.1 MPa leads to a poorly crystalline WS_2 phase, whereas subsequent sulfidation at 4.0 MPa results in the development of better-defined WS_2 structures.

8.1 Introduction

The type of catalysts used for hydrotreating processes is mainly dependent on the specific reaction and process requirements. In order to reach the more stringent fuel specifications, the removal of alkylated dibenzothiophenes, the most abundant refractory sulfur compounds in gas oil, needs to be achieved, and consequently better performing hydrodesulfurization (HDS) catalysts are required. NiW/Al_2O_3 catalysts having a higher hydrogenation activity than NiMo- and CoMo-based ones are promising for the deep HDS of diesel [1].

In sulfided NiW catalysts an active phase similar to that in sulfided CoMo catalysts has been proposed [2,3]. In analogy with the Co-Mo-S phase, the active phase in sulfided NiW catalysts should then consist of small Ni-sulfide particles adsorbed to the edges of WS_2 slabs. Type I and Type II Ni-W-S species were also distinguished in these catalysts [4,5], with a Type I phase obtained after low temperature sulfidation having high hydrogenation (HYD) activity and a Type II phase from high sulfidation temperatures with high HDS activity [4].

From a scientific point of view the sulfidation of NiW/Al_2O_3 catalysts is interesting since these catalysts are more difficult to sulfide than their Mo-based counterparts (W species are normally difficult to sulfide) and, hence, intermediate stages of sulfidation can be studied as separate phases [6]. The sulfidation degree of the W atoms dependent on the preparation procedure and the sulfidation conditions strongly influences the catalytic performance of NiW catalysts [7,8]. Therefore, the catalysts are often studied during stepwise sulfidation at different conditions to follow the formation of the active phase from the oxidic precursors.

High catalytic performance was reported for NiW/ASA (amorphous-silica-alumina) in the HDS of DBT-type molecules, making it an interesting candidate for deep-HDS applications [9]. The NiW/ASA catalysts are also excellent for first-stage hydrocracking [27]. The ASA support consists of Al_2O_3 and SiO_2, the tungstate preferentially adsorbing on the Al_2O_3 part [10]. NiW/ASA is a better catalyst than NiW/Al_2O_3 for the HDS of substituted DBT because of its higher hydrogenation ability, and also its H2S tolerance is much higher [11].

As the active sites of the catalysts are in a dynamic equilibrium with their environment, it is important to perform characterisation studies under conditions (temperature, pressure) that resemble those in the refinery [12]. Mössbauer spectroscopy is one of the very few techniques that provide the possibility to perform characterisation studies on such conditions. During the last three decades Mössbauer spectroscopy has played an important role in the characterization of hydrodesulfurization catalysts [13]. Most studies have been performed at ambient conditions while industrially sulfidation at elevated pressures is preferred. Also, the use of Mössbauer spectroscopy for studying NiW catalysts was restricted to the study of the sulfidation of Ni atoms by doping the NiW catalysts with ^{57}Co or ^{57}Fe [14].

Lee *et al.* first observed the Mössbauer effect in tungsten in 1959 [15]. Since that time only a limited number of ^{182}W Mössbauer experiments have been performed, mainly due to the special requirements that are put on the laboratory where the measurements have to be performed. For example, due to the high energy of the ^{182}W Mössbauer transition, cooling both the source and the absorber to liquid helium temperature is required to obtain a measurable resonant absorption. Some ^{182}W Mössbauer absorption spectroscopy experiments were carried out previously using standard absorbers like WSe_2 [16], WO_3 and other tungstates [17-19], WS_2 [17], WS_3, W_2C, and W_2N [20]. Here, the first ^{182}W Mössbauer study of NiW catalysts is presented. The application of ^{182}W Mössbauer spectroscopy for characterisation of catalysts is new in the world.

8.2 Experimental

Seven alumina-supported catalysts (γ-Al_2O_3, Ketjen CK300, BET surface area 263 m^2 g^{-1}, pore volume 0.66 ml g^{-1}, particle size 0.125-0.250 mm, calcined for 2 h at 573 K), prepared in earlier studies [8,14] were used as received. The catalysts were prepared by pore volume co-impregnation with aqueous solutions of nickel nitrate $Ni(NO_3)_2{\cdot}6H_2O$ (Aldrich p.a.) and ammonium metatungstate $(NH_4)_6W_{12}O_{39}{\cdot}xH_2O$ (Aldrich p.a.). The tungstate concentration was chosen such to result in a final catalyst loading (L) of 15.2 wt.%. All catalysts were dried in static air at 383 K overnight and subsequently calcined at 673 or 823 K for 2h. The various catalysts are denoted by (Ni)W(L)/ Al_2O_3 - T_c, where T_c refers to the calcination temperature. If no calcination treatment was applied, the drying temperature is indicated. The Ni/W atomic ratio employed is also given in the catalyst denotation.

Two ASA-supported samples calcined at 723 K, having 19 wt.% W and a Ni/W atomic ratio of about 0.6 at/at, were obtained from Shell (SRTCA, Amsterdam). One ASA support is a more acidic variation of the other and is denoted as ASA+.

The catalysts were sulfided in a flow of 60 cm^3 min^{-1} of 10% H_2S/H_2 mixture at pressures of 0.1 or 4 MPa in a high-pressure Mössbauer *in-situ* cell, described in section 2.1.4. The samples were heated at a rate of 6 K min^{-1} to the indicated temperature followed by an isothermal period of 2 h. The applied sulfidation treatment is denoted (S, x MPa, y K), indicating that during the experiment the catalyst is linearly heated to y K at x MPa. The experimental method applied for the ^{182}W MAS measurements is also presented in section 2.1.4.

8.3 Results and discussion

Figure 8.1 shows the ^{182}W MAS spectra for the NiW(15.2)/Al_2O_3-673 Ni:W=0.25 catalyst sulfided at atmospheric pressure and then subjected to a high-pressure treatment. The corresponding fit parameters are presented in Table 8.1. The spectra obtained with WO_3 (Aldrich p.a.) and WS_2 (Aldrich p.a.) are also presented.

The tungsten atoms in WO_3 are in a distorted octahedral environment (ReO_3 type [16]), implying an irregular coordination of oxygen to the tungsten atoms reflected by the large Q.S. and the intermediate asymmetry parameter (η) observed [20]. The measured values are similar to the ones reported previously [18, 19]. Because of the relatively short lifetime of the 100 keV transition of ^{182}W, resulting in a broad natural line width ($\Gamma \sim 2.0$ mm s^{-1}), accurate determination of isomer shifts is difficult [19].

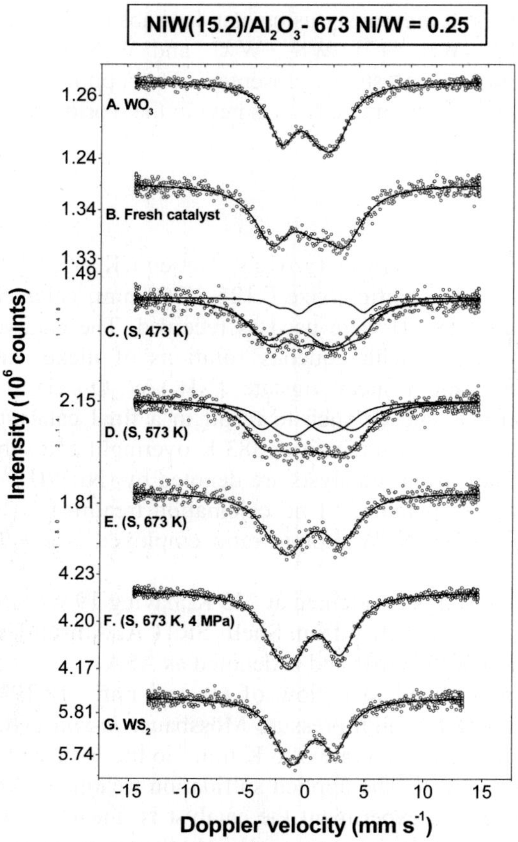

Figure 8.1 *^{182}W Mössbauer spectra of WO_3, WS_2 and NiW(15.2)/Al_2O_3-673 Ni/W = 0.25 after various successive sulfidation steps.*

Table 8.1 *^{182}W Mössbauer parameters of NiW(15.2)/Al_2O_3-673 Ni:W=0.25 catalyst after sulfidation treatment*

T_s K	P MPa	QS[a] mm s^{-1}	Γ[b] mm s^{-1}	η[c]	A[d] %	QS mm s^{-1}	Γ mm s^{-1}	η	A %
		"W – oxide"				WS_2			
WO_3		-8.4	2.6	0.5	100				
Fresh		-11	3.3	0.8	100				
673	0.1					10.5	3.2	0	100
673	4					10.1	2.8	0	100
WS_2						9.1	2.5	0	100

[a] Quadrupole splitting: Q.S. ± 0.2 mm s^{-1}.
[b] Line width: Γ ± 0.4 mm s^{-1}, [c] Asymmetry parameter: η, [d] Spectral contribution: A ± 15%.

The spectrum of the fresh catalyst resembles that of WO_3, indicating the presence of an oxidic tungsten phase in this sample. The large Q.S. and η values measured point to the existence of disordered W-oxide species in the calcined sample, which should be related to the presence of a strong W-support interaction [6]. Also the broad line width obtained indicates a distribution in local surroundings of the W atoms. Qualitatively, it can be observed that the sulfidation of tungsten starts already after exposing the catalyst to the H_2S/H_2 gas mixture at 473 K and atmospheric pressure. The presence of partially sulfided W phases (WO_xS_y) with structural parameters of disordered WO_3 species and WS_3- and WS_2-type phases at the intermediate temperatures was previously proposed [21]. Because of the occurrence of such ill-defined WO_xS_y species, having very large line widths, a quantitative analysis of the Mössbauer spectra obtained at 473 K and 573 K (Figure 8.1, C and D) is not possible. For the same reason, for all the other measured samples, only the spectra obtained after sulfidation at 673 K will be presented.

After treatment at 673 K the spectral shape is similar to that of WS_2 (Figure 8.1, E), but the large Q.S. and line width values indicate that only a poorly crystalline WS_2 phase is present after sulfidation at 0.1 MPa. The structure of WS_2 is similar to that of MoS_2 in which the Mo atoms are in the center of a regular trigonal prism – a structure that is in accordance with the observation of an axially symmetric electric field gradient ($\eta = 0$) [19]. Upon increasing the sulfidation pressure to 4 MPa at 673 K (Figure 8.1, F), the WS_2 slabs in NiW(15.2)/Al_2O_3-673 Ni:W=0.25 catalyst become better crystallized (as resulted from the small line width measured). For this sample, an increased thiophene HDS activity was observed upon increasing the sulfidation pressure from 0.1 to 1.5 MPa [8].

Table 8.2 *^{182}W Mössbauer parameters of W(15.2)/Al_2O_3-673 catalyst after sulfidation treatment*

T_s K	P MPa	QS mm s^{-1}	Γ mm s^{-1}	η	A %	QS mm s^{-1}	Γ mm s^{-1}	η	A %
			"W – oxide"				WS_2		
Fresh		-10.9	3.8	0.8	100				
			WS_3						
673	0.1	-13.8	3.2[a]	0	61	10.4	3.2[a]	0	39
673	4					9.8	3.0	0	100

[a] Fixed during fit.

The Mössbauer spectra of the W(15.2)/Al_2O_3-673 catalyst sulfided at 0.1 and 4 MPa are presented in Figure 8.2 and the resulting MES parameters are shown in Table 8.2. The spectrum of the fresh catalyst shows a very broad line width, pointing to the presence of less defined W-oxide species as compared to the similar calcined sample containing also Ni. After treatment at 673 K, the presence of a quadrupole coupling of the same sign as WO_3 but with a zero asymmetry parameter is observed – this component being assigned to WS_3 [20]. The large Q.S. value indicates the presence of a very disordered WS_3-type structure. A proposal for intermediate WS_3 species in the sulfidation of NiW/Al_2O_3 and supported WO_3 has also been made earlier [9, 22]. The partial formation of the poorly crystalline WS_2 phase is also observed at this point. After sulfidation at 673 K and 4 MPa, the W sulfidation is complete, although the WS_2 species are still less well defined (large line width observed).

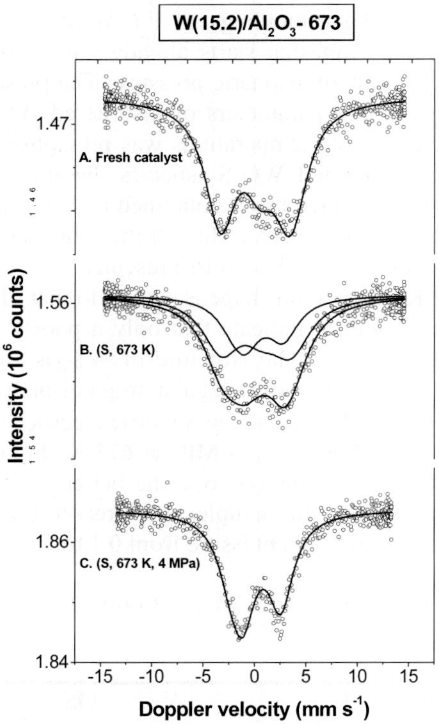

Figure 8.2 *^{182}W Mössbauer spectra of W(15.2)/Al_2O_3-673 catalyst – after various successive sulfidation steps.*

Figure 8.3 shows the MAS spectra of NiW(15.2)/Al_2O_3-823 Ni:W=0.25 sulfided at atmospheric pressure and subsequently at high-pressure. The corresponding fit parameters are presented in Table 8.3.

Table 8.3 *^{182}W Mössbauer parameters of NiW(15.2)/Al_2O_3-823 Ni:W=0.25 catalyst after sulfidation treatment*

T_s K	P MPa	QS mm s^{-1}	Γ mm s^{-1}	η	A %	QS mm s^{-1}	Γ mm s^{-1}	η	A %
		"W – oxide"				WS_2			
Fresh		-12.1	3.6	0.8	100				
673	0.1					10.9	3.7	0	100
673	4					10.4	3.0	0	100

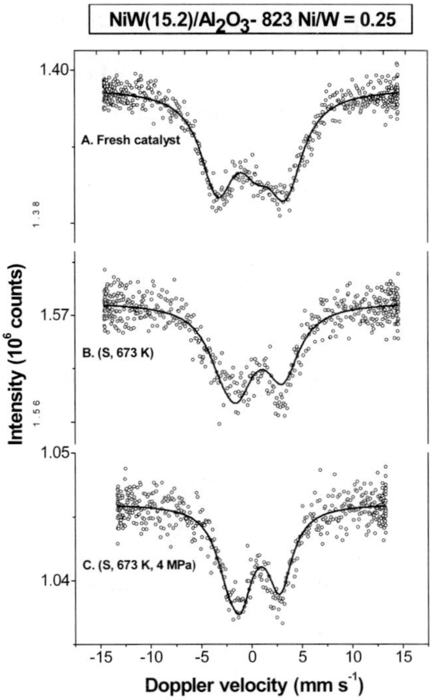

Figure 8.3 *^{182}W Mössbauer spectra of NiW(15.2)/Al_2O_3-823 Ni/W = 0.25 – after various successive sulfidation steps.*

The large Q.S. value observed for the oxidic tungsten phase in the fresh catalyst indicates a more irregular coordination of oxygen to the tungsten atoms. This can be understood in terms of a stronger W-support interaction, with more W-O-Al bonds following the calcination treatment at an elevated temperature. Complete transition to poorly crystalline WS_2 phase takes place after sulfidation at 673 K and atmospheric pressure, the WS_2 crystallization degree being smaller compared to that of the sample calcined at 673 K treated in the same conditions. After sulfidation at 4.0 MPa (Figure 8.3, C), the WS_2 structures are better defined, but not to the same extent as seen in the high-pressure sulfided NiW(15.2)/Al_2O_3-673 Ni:W=0.25 sample. The higher calcination temperature is known to induce a more difficult sulfidation of W atoms and this leads to lower thiophene HDS activities [8, 23, 24].

The Mössbauer spectra of W(15.2)/Al_2O_3-823 catalyst sulfided at 673 K are presented in Figure 8.4 and the resulting MAS parameters are shown in Table 8.4. The spectrum of the fresh catalyst was fitted with the same large Q.S. contribution as observed for its Ni-containing counterpart, confirming the presence of very disordered W-oxide species upon calcination at 823 K. After atmospheric-pressure sulfidation at 673 K, the formation of the disordered WS_3-type species is found. The low sulfidation rate of supported WO_3 materials allows the observation of WS_3 even at this high sulfidation temperature. This suggests that the presence of Ni modifies to some extent the sulfidation mechanism. Tentatively, the Ni-sulfide species in NiW/Al_2O_3 catalysts stabilize intermediate W species. The final W sulfidation degree of Ni-containing samples is higher. Upon treatment at 673 K and 4 MPa, the W(15.2)/Al_2O_3-823 catalyst appears to be better sulfided, although the formed WS_2 species are again poorly crystallized.

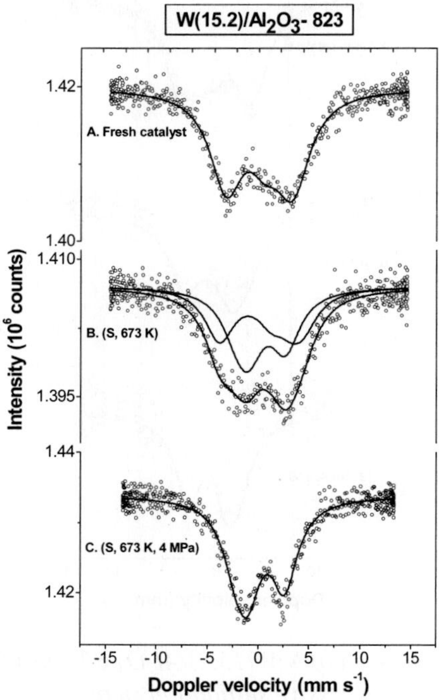

Figure 8.4 *^{182}W Mössbauer spectra of W(15.2)/Al_2O_3-823 – after various successive sulfidation steps.*

Table 8.4 *^{182}W Mössbauer parameters of W(15.2)/Al_2O_3-823 catalyst after sulfidation treatment*

T_s K	P MPa	QS mm s^{-1}	Γ mm s^{-1}	η	A %	QS mm s^{-1}	Γ mm s^{-1}	η	A %
			"W – oxide"				WS_2		
Fresh		-11.8	3.6	0.8	100				
			WS_3						
673	0.1	-13.5	3.2[a]	0	45	11.0	3.2[a]	0	55
673	4					9.8	3.2	0	100

[a] Fixed during fit.

Figure 8.5 shows the ^{182}W MAS spectra obtained with NiW(15.2)/Al_2O_3-383 Ni:W=0.25 catalyst sulfided at 0.1 and 4 MPa and the corresponding fit parameters are collected in Table 8.5.

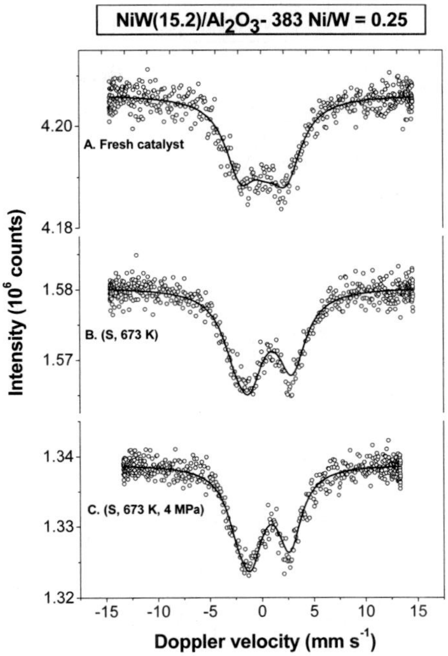

Figure 8.5 *^{182}W Mössbauer spectra of NiW(15.2)/Al_2O_3-383 Ni/W = 0.25 – after various successive sulfidation steps.*

Table 8.5 *^{182}W Mössbauer parameters of NiW(15.2)/Al_2O_3-383 Ni:W=0.25 catalyst after sulfidation treatment*

T_s K	P MPa	QS mm s^{-1}	Γ mm s^{-1}	η	A %	QS mm s^{-1}	Γ mm s^{-1}	η	A %
			"W – oxide"				WS_2		
Fresh		-7.7	3.2	0.9	100				
673	0.1					10.0	3.0	0	100
673	4					9.9	2.8	0	100

The spectrum of the fresh catalyst shows the presence of a contribution with an unexpectedly low Q.S. value, which should probably be assigned to a Ni-W-O interaction phase. The sulfidation of W atoms is already complete after treatment at 673 K and 0.1 MPa, the resulted WS_2 structures being better defined as compared to those present in the similar sample calcined at 673 K and sulfided in the same conditions. The NiW(15.2)/Al_2O_3-383 Ni:W=0.25 catalyst activated under atmospheric pressure conditions has higher thiophene HDS activities than the comparable calcined catalysts [25]. After treatment at 673 K and 4 MPa, better-crystallized WS_2 species are formed.

The Mössbauer spectra of the W(15.2)/Al_2O_3-383 sample sulfided at 673 K are presented in Figure 8.6 and the resulting MAS parameters are shown in Table 8.6.

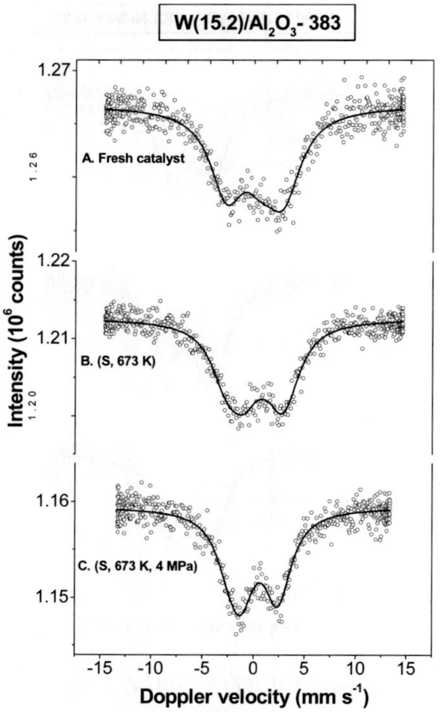

Figure 8.6 *[182]W Mössbauer spectra of W(15.2)/Al_2O_3-383 – after various successive sulfidation steps.*

Table 8.6 *[182]W Mössbauer parameters of W(15.2)/Al_2O_3-383 catalyst after sulfidation treatment*

T_s K	P MPa	QS mm s^{-1}	Γ mm s^{-1}	η	A %	QS mm s^{-1}	Γ mm s^{-1}	η	A %
		"W – oxide"				WS_2			
Fresh		-10.2	3.7	0.7	100				
673	0.1					9.8	3.8	0	100
673	4					9.6	3.2	0	100

The spectrum of the fresh catalyst shows the presence of a somewhat more regular W-oxide structure as compared with the previous samples, although the broad line width still points to a large distribution in local surroundings of the W atoms. Upon treatment at 673 K under both atmospheric and high-pressure conditions, the sulfidation of the W(15.2)/Al_2O_3-383 samples appears to be complete, but the resulted WS_2 slabs are clearly poorly crystallized, as observed from the large line widths measured.

Figure 8.7 shows the [182]W MAS spectra obtained with NiW(15.2)/Al_2O_3-673 Ni:W=0.6 catalyst sulfided at 673 K and the corresponding fit parameters are collected in Table 8.7.

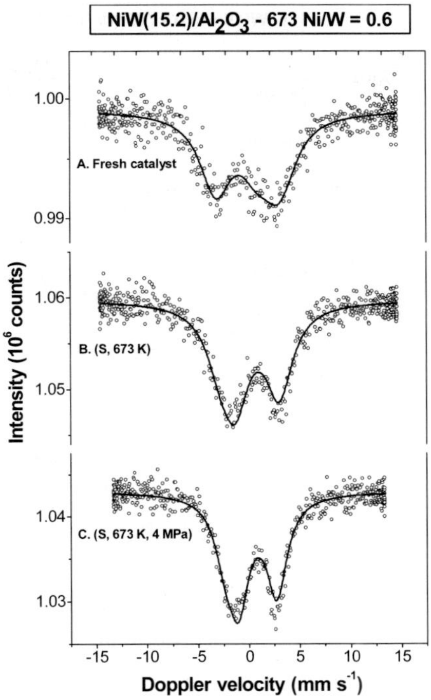

Figure 8.7 *^{182}W Mössbauer spectra of NiW(15.2)/Al$_2$O$_3$-673 Ni/W = 0.6 – after various successive sulfidation steps.*

Table 8.7 *^{182}W Mössbauer parameters of NiW(15.2)/Al$_2$O$_3$-673 Ni:W=0.60 catalyst after sulfidation treatment*

T_s K	P MPa	QS mm s^{-1}	Γ mm s^{-1}	η	A %	QS mm s^{-1}	Γ mm s^{-1}	η	A %
		"W – oxide"				WS_2			
Fresh		-11.7	3.4	0.6	100				
673	0.1					10.5	3.0	0	100
673	4					10.1	2.5	0	100

The sulfidation pattern of the W atoms is similar for NiW(15.2)/Al$_2$O$_3$-673 Ni:W=0.6 and NiW(15.2)/Al$_2$O$_3$-673 Ni:W=0.25 catalysts (Tables 8.7 and 8.1) after activation at both 0.1 and 4 MPa. However, the WS_2 species are less crystallized in the high promoter-loading catalyst after the 0.1 MPa treatment, leading to a higher thiophene HDS activity compared to the Ni:W=0.25 sample [14]. Complete crystallization of the WS_2 slabs is observed with the NiW(15.2)/Al$_2$O$_3$-673 Ni:W=0.6 catalyst upon sulfidation at 673 K and 4 MPa, the measured line width being similar to that of bulk WS_2 (Table 8.1). The more crystalline WS_2 slabs present low dispersion and low intrinsic activities [8].

The Mössbauer spectra of the NiW(19)/ASA-723 Ni:W=0.6 catalyst sulfided at 0.1 and 4 MPa are presented in Figure 8.8 and the resulting MES parameters are shown in Table 8.8.

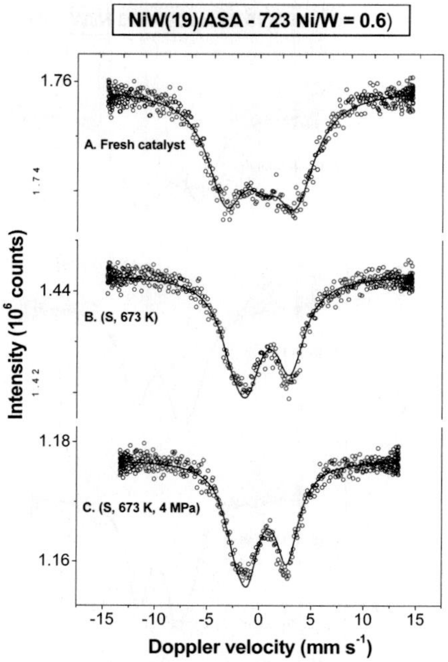

Figure 8.8 *^{182}W Mössbauer spectra of NiW(19)/ASA-723 Ni:W=0.6 – after various successive sulfidation steps.*

Table 8.8 *^{182}W Mössbauer parameters of NiW(19)/ASA-723 Ni:W=0.6 catalyst after sulfidation treatment*

T_s K	P MPa	QS mm s^{-1}	Γ mm s^{-1}	η	A %	QS mm s^{-1}	Γ mm s^{-1}	η	A %
		"W – oxide"				WS_2			
Fresh		-11.7	3.9	0.9	100				
673	0.1					10.1	3.1	0	100
673	4					10.1	2.9	0	100

Figure 8.9 shows the Mössbauer spectra obtained with the NiW(19)/ASA+ -723 Ni:W=0.6 catalyst sulfided at 673 K and the corresponding fit parameters are collected in Table 8.9. The sulfidation behavior of the ASA-supported NiW catalysts is very similar to the alumina-supported ones, but the WS_2 crystallization degree appears to be higher for the NiW(19)/ASA+-723 Ni:W=0.6 catalyst after activation at both 0.1 and 4 MPa. However, no clear indication of any electronic effect has emerged. Trendwise, there is not much difference in spectral parameters with the alumina-supported ones (Figure 8.7) and this is no doubt due to the preference of tungsten to be located on the alumina part of the ASA support. The higher DBT HDS activities obtained with NiW catalysts supported on ASA+ were previously related to an increased hydrogenation (HYD) activity [11]. The HYD rate is believed to increase with an increasing stacking degree because of a more favorable planar adsorption geometry of reactants on multilayered Mo(W)S_2 [26].

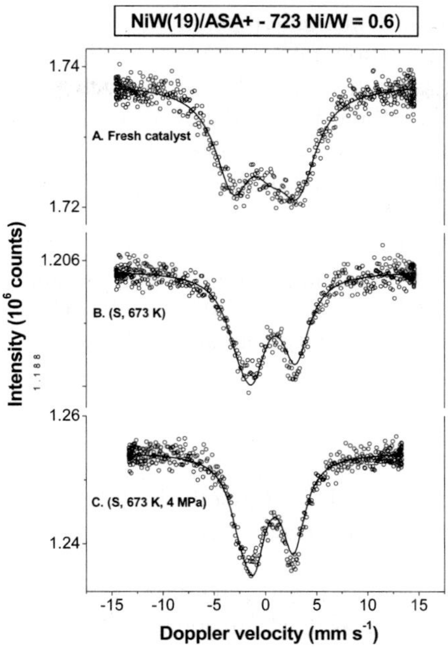

Figure 8.9 *^{182}W Mössbauer spectra of NiW(19)/ASA+ -723 Ni:W=0.6 – after various successive sulfidation steps.*

Table 8.9 ***^{182}W*** *Mössbauer parameters of NiW(19)/ASA+ -723 Ni:W=0.6 catalyst after sulfidation treatment*

T_s K	P MPa	QS mm s^{-1}	Γ mm s^{-1}	η	A %	QS mm s^{-1}	Γ mm s^{-1}	η	A %
		"W – oxide"				WS_2			
Fresh		-11.4	3.8	0.8	100				
673	0.1					10.2	2.9	0	100
673	4					10.3	2.7	0	100

8.4 Conclusions

The sulfidation of Al_2O_3- and ASA- supported NiW catalysts under conditions relevant to industrial practice was studied by ^{182}W Mössbauer spectroscopy. This is the first study of its kind employing ^{182}W Mössbauer absorption spectroscopy for characterisation of catalysts.

^{182}W MAS of supported tungsten catalysts can distinguish between WO_3- and WS_2-type phases as encountered in the calcined and sulfided forms of W-based hydrotreating catalysts. In general, the line widths are substantially broadened compared to crystalline WO_3 and WS_2 reference compounds because small, disordered particles interacting with the support are present in such catalysts. Some indications for the presence of oxysulfidic W or WS_3 intermediates have been found. It is typically observed that enhanced support interaction through increased calcination temperatures leads to more difficult sulfidation of W. In cases where the sulfidation rate is lowest, an intermediate WS_3-type phase is identified at temperatures as high as 673 K. Ni facilitates sulfidation and in this case sulfidation occurs via an oxysulfidic intermediate. A general conclusion is that sulfidation at 673 K and 0.1 MPa leads to a poorly crystalline WS_2 phase. Subsequent sulfidation at 4.0 MPa increases the crystallinity of the catalysts as evidenced by the approach of the spectral parameters to those of WS_2.

^{182}W MAS can be a useful tool to study W-based hydrotreating catalysts. Given the more difficult sulfidability of W, combination of this technique with structural analysis around W (EXAFS) and catalytic activity studies might be very promising to study the difficult issue of phase I/II behavior in hydrotreating catalysts.

References

1. H. Topsøe, B.S. Clausen, F.E. Massoth, in: J.R. Anderson, M. Boudart (Eds.), Hydrotreating Catalysts, in: Catal. Science Technol., vol. 11, Springer, Berlin, 1996.
2. S.P.A. Louwers, R. Prins, J. Catal. 139 (1993) 525.
3. H. Shimada, N. Matsubayashi, M. Imamura, T. Sato, Y. Yoshimura, T. Kameoka, K. Masuda, A. Nishijima, Bull. Soc. Chim. Belg. 104 (1995) 353.
4. M. Breysse, J. Bachelier, J.P. Bonnelle, M. Cattenot, D. Cornet, T. Décamp, J.C. Duchet, R. Durand, P. Engelhard, R. Frety, C. Gachet, P. Geneste, J. Grimblot, C. Gueguen, S. Kasztelan, M. Lacroix, J.C. Lavalley, C. Leclercq, C. Moreau, L. De Mourgues, J.L. Olivé, E. Payen, J.L. Portefaix, H. Toulhoat, M. Vrinat, Bull. Soc. Chim. Belg. 96 (1987) 829.
5. H.R. Reinhoudt, C.H.M. Boons, A.D. van Langeveld, J.A.R. van Veen, S.T. Sie, J.A. Moulijn, Appl. Cat. A 207 (2001) 25.
6. B. Scheffer, P.J. Mangnus, J.A. Moulijn, J. Catal. 121 (1990) 18.
7. H. R. Reinhoudt, Y. van der Meer, A. M. van der Kraan, A. D. van Langeveld and J. A. Moulijn, Fuel Processing Technology, 61 (1999) 43.
8. E.J.M. Hensen, Y. van der Meer, J.A.R. van Veen, J.W. Niemantsverdriet, Appl. Catal. A 322 (2007) 16.
9. H. R. Reinhoudt, R. Troost, A. D. van Langeveld, S.T. Sie, J.A.R. van Veen, J. A. Moulijn, Fuel Processing Technology, 61 (1999) 133.
10. M.J. Vissenberg, L.J.M. Joosten, M.E.H. Heffels, A.J. van Welsenes, V.H.J. de Beer, R.A. van Santen, J.A.R. van Veen, J. Phys. Chem. B 104 (2000) 8456.
11. W.R.A.M. Robinson, J.A.R. van Veen, V.H.J. de Beer, R.A. van Santen, Fuel Processing Technology, 61 (1999) 103.
12. H. Topsøe, J. Catal. 216 (2003) 155.
13. M.W.J. Crajé, V.H.J. de Beer, J.A.R. van Veen, A.M. van der Kraan, in: M.L. Occelli, R.R. Chianelli (Eds.), Hydrotreating Technology for Pollution Control, Dekker, New York, 1996, p. 95.
14. Y. van der Meer, PhD thesis, Delft University of Technology, Delft, 2001, ISBN 90-407-2230-X.
15. L. Lee, L. Meyer-Schutzmeister, J. Schiffer, D. Vincent, Phys. Rev. Letters 13 (1959) 223.

16. M.G. Clark, R. Gancedo, A.G. Maddock, A.F. Williams, A.D. Yoffe, J. Phys. C 6 (1973) 474.
17. N.N. Savvateev, K.V. Pokholok, A.M. Babechkin, B.E. Dzevitsky, G.N. Zviadadze, Solid State Communications 39 (1981) 793.
18. D. Agresti, E. Kankeleit, B. Persson, Phys. Rev. 155 (1969) 1342.
19. G.M. Bancroft, R.E.B. Garrod, A.G. Maddock, Inorg. Nucl. Chem. Letters 7 (1971) 1157.
20. A.G. Maddock, R.H. Platt, A.F. Williams, R. Gancedo, J Chem. Soc. Dalton Trans. 12 (1974) 1314.
21. Y. van der Meer, E.J.M. Hensen, J.A.R. van Veen and A.M. van der Kraan, J. Catal. 228 (2004) 433.
22. M. Sun, T. Bürgi, R. Cattaneo, A.D. van Langeveld, R. Prins J. Catal. 201 (2001) 258.
23. J.C. Duchet, J.C. Lavalley, S. Housni, D. Ouafi, J. Bachelier, M. Lakhdar, A. Mennour, D. Cornet, Catal. Today 4 (1988) 71.
24. H.R. Reinhoudt, E. Crezee, A.D. van Langeveld, P.J. Kooyman, J.A.R. van Veen, J. A. Moulijn, J. Catal. 196 (2000) 315.
25. M.J. Vissenberg, Y. van der Meer, E.J.M. Hensen, V.H.J. de Beer, A.M. van der Kraan, R.A. van Santen, J.A.R. van Veen, J. Catal. 198 (2001) 151.
26. E.J.M. Hensen, P.J. Kooyman, Y. van der Meer, A.M. van der Kraan, V.H.J. de Beer, J.A.R. van Veen, R.A. van Santen, J. Catal. 199 (2001) 224.
27. J.K. Minderhoud, J.A.R. van Veen, Fuel Proc. Technol. 35 (1993) 87.

16. M.G. Clark, [illegible], A.G. Maddock, A.F. Williams, A.D. Yoffe, J. Phys. C 6 (1973) 474.
17. N. N. [illegible], [illegible], B. [illegible], G.N. Zviadadze, Solid State Communications 39 (1981) 701.
18. [illegible], Phys. Rev. 155 (1967) 1242.
19. G.M. Bancroft, [illegible], A.G. Maddock, [illegible] Chem. Letters 7 (1971) 1157.
20. A.G. Maddock, [illegible], A.F. Williams, R. Glasscoe, J. Chem. Soc. Dalton Trans. 12 (1973) [illegible]
21. [illegible] van der [illegible], [illegible] Hensen, J.A.R. van Veen, and A.M. van der Kraan, J. Catal. 225 (2004) [illegible]
22. M. Sun, [illegible], A.D. van Langeveld, R. Prins, J. Catal. 201 (2001) 258.
23. [illegible], D. Ouafi, [illegible], M. Lakhdar, A. [illegible], Catal. Today 4 (1989) 71.
24. [illegible], A.D. van Langeveld, P.J. Kooyman, J.A.R. van Veen, J.A. Moulijn, J. Catal. 196 (2000) 315.
25. [illegible] de Beer, A.M. van der Kraan, [illegible], J.A.R. van Veen, J. Catal. 198 (2001) 151.
26. E.J.M. Hensen, [illegible], [illegible] van der Meer, A.M. van der Kraan, V.H.J. de Beer, J.A.R. van Veen, R.A. van Santen, J. Catal. 199 (2001) 224.
27. [illegible], J.A.R. van Veen, Fuel Proc. Technol. 35 (1993) 87.

Summary

High-Pressure Sulfidation of Hydrotreating Catalysts:
Genesis and Properties of the Active Phase

PhD Thesis by A.I. Dugulan

The global petroleum demand is projected to increase in the next decennia as oil is expected to remain the primary source of energy around the globe contributing to approximately 40% of the total consumption. The production of high-quality fuels from crude oil involves physical separation steps but also catalytic processes to remove contaminants (hydrotreating), shift the boiling point (alkylation and hydrocracking) and increase fuel quality (isomerisation). In hydrotreating sulfur, nitrogen, oxygen and metal atoms are removed from the different petroleum streams and unsaturated hydrocarbons are hydrogenated. The main reasons for refineries to perform hydrotreating are of environmental and economic nature. Besides protection of downstream catalysts from poisoning by sulfur, stringent environmental legislation has been aimed at the reduction of sulfur oxide emissions from fuel combustion. The environmental regulations on transportation fuel quality and the diminishing supplies of lighter types of crude oil necessitate further improvement of hydrotreating processes. An important approach is then to further optimize the catalytic activity of hydrotreating catalysts.

Hydrodesulfurization (HDS) catalysts consist of mixed sulfides of Co or Ni and Mo or W typically dispersed over a γ-alumina support. The choice for metal combination depends mainly on the application and the desired activity and selectivity of the catalysts. CoMo sulfide catalysts are preferred for HDS operations, while NiMo sulfides are better suitable when hydrodenitrogenation is important and when deeper hydrogenation is required. NiW sulfide catalysts have the highest activity for hydrogenation, but a drawback is the higher catalyst cost due to the use of tungsten. A generally accepted model of the active phase of CoMo sulfide catalysts is one in which Co atoms are located at the edges of MoS_2 slabs. This phase is denoted as the Co-Mo-S phase. The occurrence of two different Co-Mo-S structures is assumed in alumina-supported catalysts: an incompletely sulfided Type I and a more active Type II Co-Mo-S, where all Mo-O-Al linkages with the alumina support are sulfided.

In spite of the substantial progress made in the fundamental understanding of the active phase morphology, the metal-support interaction and reaction mechanisms, many details about the nature and stability of the active sites have not yet been elucidated. One specific issue not widely addressed is the influence of the sulfidation pressure on the active phase in hydrotreating catalysts. In industrial practice, catalysts are brought in their active, sulfided form at elevated pressure. The main objective of the present study was to understand the effect of the sulfidation pressure on the active phase structure in CoMo and NiW catalysts.

The influence of the sulfidation pressure on the active phase structure in CoMo-based hydrotreating catalysts was investigated by ^{57}Co Mössbauer emission spectroscopy (MES). Here, the first in situ high-pressure MES study of the sulfidation of CoMo hydrodesulfurization catalysts is reported. ^{182}W Mössbauer absorption spectroscopy (MAS) for W-containing solids has been implemented and the first-time report of ^{182}W Mössbauer spectra on the sulfidation of supported (Ni)W catalysts is presented. Detailed structural information about the catalysts under industrially relevant conditions can be obtained using Mössbauer spectroscopy as a fingerprint technique. Complementary structural information has been obtained by extended X-ray absorption fine structure (EXAFS) and transmission electron microscopy (TEM) measurements. An important step in the formulation of design criteria is the establishment of structure-activity correlations which aims to relate catalytic performance to specific catalyst properties. To this end, the catalytic activity in the gas-phase hydrodesulfurization (HDS) of dibenzothiophene was determined.

In chapter 3, the influence of the sulfidation pressure on the structure and activity of calcined $CoMo/Al_2O_3$ catalysts was investigated. The amount of promoter atoms that participate in the active phase is higher after sulfidation at 4 MPa than after atmospheric pressure sulfidation. A study of the stepwise sulfidation of such samples points to higher sulfidation rates of Co and Mo at elevated pressure. This leads to somewhat more aggregated CoS_x species at intermediate temperatures, followed by redispersion of these species to more dispersed ones. This redispersion occurs at the moment when the MoS_2 slabs are formed. Furthermore, it is shown that a typical calcined $CoMo/Al_2O_3$ sample, which expectedly behaves as a Type I catalyst following atmospheric pressure sulfidation, renders a much more active Type II catalyst upon high-pressure activation. This observation calls for a closer look at the question to what extent it is possible to preserve Type I Co-Mo-S under industrial sulfidation conditions.

To better understand the influence of calcination conditions on the performance of $CoMo/Al_2O_3$ catalysts, in chapter 4 we investigated a set of uncalcined catalysts with different Co/Mo ratios. Omitting calcination after loading the calcined Mo/Al_2O_3 catalyst with Co leads to a decreased interaction of the promoter ions with the Al_2O_3 support. As a consequence, all promoter atoms are completely sulfided at lower temperatures. However, a significant fraction of small CoS_x particles or highly dispersed Co_9S_8-type species with a low activity form when the Co/Mo ratio is relatively high. Although the CoS_x structures can have MES parameters that resemble those of the Co-Mo-S phase, MES can still indicate the presence of different Co-sulfide species, since smaller quadrupole splitting (Q.S.) values are clearly pointing to the occurrence of larger Co sulfide species. The high-pressure sulfided samples have a much higher HDS activity than the samples activated at 0.1 MPa. The difference is due to formation of Type II Co-Mo-S phase at elevated pressures. The activities of the samples with Co/Mo ratios of 0.30 and 0.52 are very similar which further underpins the finding that for the catalyst with the higher promoter content a substantial part of the Co has ended up in segregated CoS_x structures.

Chapter 5 evaluates the applicability of the Co-Mo-S structural model and the stability of the Co-containing phases for carbon-supported catalysts activated under high-pressure conditions. The weak interaction of the promoter atoms with the support and also the weak Co-Mo interaction leads to a significant formation of more aggregated Co-sulfide structures after sulfidation at 0.1 MPa. MoS_2 hinders the complete sintering of the CoS_x structures under atmospheric pressure conditions, but upon more severe sulfidation treatments at 4 MPa these CoS_x species sinter to Co_9S_8-type structures. Atmospheric pressure sulfidation results in active species, likely highly dispersed Type II Co-Mo-S particles, constrained in the micropores of the carbon support. The lower accessibility has a negative effect on the DBT HDS activity but the thiophene HDS activity of such a catalyst is relatively higher. On the contrary, sulfidation at 4 MPa leads to sintering of the MoS_2 phase as also follows from the lower thiophene HDS activity. The DBT HDS activity is now much higher because the particles are not constrained anymore in the micropores.

To better understand the influence of high-pressure sulfidation of $CoMo/Al_2O_3$ catalysts prepared via non-conventional routes, nitrilotriacetic acid (NTA) and phosphorus containing CoMo catalysts were investigated in chapter 6. The presence of additives, which decrease the metal-support interaction, leads to a significant formation of CoS_x species. Only the Co-Mo-S structures are present in the lower-loading catalysts, but an improved dispersion of the Co atoms in the sample prepared using larger amount of NTA was noticed. The spectroscopic data suggest that a complete formation of Type II Co-Mo-S species takes place only after the high-pressure treatment. Higher activities are obtained with catalysts treated also at 4 MPa compared to the samples that are activated only at atmospheric pressure.

In chapter 7, an attempt is made to understand the strongly varying DBT HDS activities among the various catalysts presented in earlier chapters. Since our spectroscopic results probing the environment around Co (MES, EXAFS) did not reveal any significant differences in the final state, the attention was focused on the Type I/II behavior. The existence of a continuum of Co-Mo-S structures ranging from the less active Type I phase having a strong interaction with the support up to a fully sulfided and possibly well-crystallized Type II Co-Mo-S phase with a weak support interaction is formulated for sulfided CoMo catalysts. The use of NTA and high-pressure sulfidation leads to the highest Type II character of the active Co-Mo-S phase. A reasonable indicator of the Type I/II character of the active phase appears to be the Mo-S coordination number derived from Mo K-edge X-ray absorption spectroscopy. Concomitant with reaching a full sulfidation degree in NTA catalysts, stacking of the MoS_2 slabs occurs. This stacking does not appear to influence gas-phase thiophene HDS much but impedes activity in gas-phase DBT HDS to some extent. It is suggested that the DBT HDS activity of CoMo-NTA catalyst sulfided at high-pressure (pure phase II) with a somewhat stacked MoS_2 morphology may be improved by limiting the stacking of the active phase. The most active catalyst in DBT HDS is a calcined CoMo catalyst sulfided at 4 MPa – the high activity is explained by nearly complete sulfidation to phase II whilst keeping an almost monolayer MoS_2 morphology. The requirements of a low stacking degree and full MoS_2 sulfidation are discussed for highly active CoMo sulfide catalysts.

Chapter 8 describes the sulfidation of Al_2O_3- and amorphous-silica-alumina-supported NiW catalysts under conditions relevant to industrial practice using ^{182}W Mössbauer spectroscopy for the first time. ^{182}W MAS can clearly distinguish between WO_3- and WS_2-type phases encountered in the calcined and sulfided forms of W-based hydrotreating catalysts. In addition to the study of the sulfidation degree of a catalyst, ^{182}W MAS also provides an estimate of the degree of crystallinity of the WS_2 species, which is relevant to the phase I/II behavior in NiW catalysts. This type of information is very useful, but, on the other hand, may also be obtained more straightforwardly with other more conventional techniques. For NiW supported on alumina and amorphous-silica alumina, it was typically observed that enhanced support interaction through increased calcination temperatures leads to more difficult sulfidation of W. In cases where the sulfidation rate is lowest, an intermediate WS_3-type phase is identified at temperatures as high as 673 K. Ni facilitates sulfidation and in this case sulfidation occurs via an oxysulfidic intermediate. A general conclusion is that sulfidation at atmospheric pressure leads to a poorly crystalline WS_2 phase, while sulfidation at 4.0 MPa increases the crystallinity of the catalysts as evidenced by the approach of the spectral parameters to those of WS_2.

In this study we have identified two major factors influencing the HDS activities of CoMo hydrotreating catalysts: the significant formation of CoS_x structures in catalysts with weak metal-support interaction and the existence of a continuum of Type I/II Co-Mo-S structures, where the fully Type II end member is more difficult to reach than previously suspected. MES was confirmed as a viable technique to at least clearly indicate the presence of the promoter atoms in Co-sulfide species involving variable number of Co atoms. What we still need is a better understanding of the stacking effect on the HDS activity and the determination of the best choice of catalyst preparation and activation methods in order to achieve full MoS_2 sulfidation with a reasonably high dispersion. As unequivocally demonstrated in this thesis, the importance of conducting the future experiments under temperature and pressure conditions relevant to industrial practice is paramount. A better quantitative description of the Co-species present in CoMo sulfide catalysts (possibly by using MES and the quadrupole splitting distribution approach) is also indicated as a potential avenue for future research.

Samenvatting

Inzwaveling onder Hoge Druk van *Hydrotreating* Katalysatoren:
Vorming en Eigenschappen van de Actieve Fase

PhD Thesis door A.I. Dugulan

Gezien de groeiende vraag naar energie ligt het in de verwachting dat de wereldwijde vraag naar aardolie zal blijven toenemen. Aardolie zal ook in de toekomst met een aandeel van 40% in de totale energievoorziening een belangrijke rol blijven spelen. The productie van brandstoffen van hoge kwaliteit uit ruwe aardolie bestaat uit fysische scheidingstechnieken, maar uit ook katalytische processen om vervuilingen te verwijderen (*hydrotreating*), het kookpunt te veranderen (alkylering en kraken) en de brandstofkwaliteit te verbeteren (isomerisatie). In *hydrotreating* worden zwavel-, stikstof-, zuurstof- en metaalatomen verwijderd uit de verschillende aardoliefracties en worden onverzadigde koolwaterstoffen gehydrogeneerd. De belangrijkste redenen voor raffinaderijen om hun producten te hydrotreaten zijn van economische als mileutechnische aard. Behalve om katalysatoren verderop in het raffinageproces te beschermen tegen zwavelverontreinigingen is er strengere milieuwetgeving van kracht die erop gericht is de uitstoot van zwaveldioxide door verbranding van brandstoffen te reduceren. De strengere milieu-eisen aan transportbrandstoffen en de afnemende voorraden van lichtere soorten van ruwe olie maken het noodzakelijk om *hydrotreating* processen verder te verbeteren. Een belangrijke aanpak is de verbetering van de katalytische activiteit van *hydrotreating* katalysatoren.

Ontzwavelingskatalysatoren bestaan uit combinaties van Co- of Ni-sulfides met Mo- of W-sulfides die meestal gedispergeerd worden op een γ-alumina dragermateriaal. De keuze van de combinatie hangt voornamelijk af van de toepassing en de gewenste activiteit en selectiviteit van de katalysator. CoMo-sulfide katalysatoren hebben de voorkeur bij ontzwaveling, terwijl NiMo-sulfides beter geschikt zijn wanneer ontstikstoffing belangrijk is en wanneer een diepere hydrogenering wordt vereist. NiW-sulfide katalysatoren hebben de hoogste activiteit voor hydrogenering, maar een nadeel zijn de hogere kosten door het gebruik van wolfraam. In het algemeen geaccepteerde Co-Mo-S model van de actieve fase in CoMo-sulfide katalysatoren zijn de Co-atomen gelokaliseerd aan de randen van MoS_2 *slabs*. Er wordt verondersteld dat er twee verschillende Co-Mo-S structuren voorkomen in alumina-gedragen katalysatoren: een onvolledig ingezwavelde Type I fase en een actievere Type II Co-Mo-S fase waarbij alle Mo-O-Al verbindingen met de alumina drager zijn ingezwaveld.

Ondanks dat er aanzienlijke vooruitgang is geboekt in het fundamentele begrip van de morfologie van de actieve fase, de metaal-drager interactie en reactiemechanismes zijn er nog veel details over de aard en de stabiliteit van de actieve fase onopgehelderd. Een specifiek aspect dat nog niet uitgebreid is beschreven is de invloed van de druk tijdens inzwaveling op de actieve fase in *hydrotreating* katalysatoren. In de industrie worden katalysatoren in hun actieve, ingezwavelde vorm gebracht bij verhoogde druk. Het belangrijkste doel van deze studie was het begrijpen van de invloed van deze inzwavelingsdruk op de actieve fase in CoMo en NiW katalysatoren.

De invloed van de inzwaveldruk op de structuur van de actieve fase in CoMo-achtige *hydrotreating* katalysatoren is voornamelijk onderzocht met ^{57}Co Mössbauer emissie spectroscopie (MES). In dit onderzoek worden de eerste hoge-druk MES resultaten van CoMo ontzwavelingskatalyatoren gepresenteerd. ^{182}W Mössbauer absorptie spectroscopie (MAS) voor W-houdende vaste stoffen is operationeel gemaakt en voor het eerst worden er

^{182}W Mössbauer spectra van de inzwaveling van gedragen (Ni)W katalysatoren gepubliceerd. Met Mössbauer spectroscopie als '*fingerprint*' techniek kan gedetailleerde informatie over de structuur van de katalysatoren onder relevante industriële condities verkregen worden. Met *X-ray absorption fine structure* (EXAFS) en transmissie electronen microscopie (TEM) is daarnaast complementaire informatie over de actieve fase verzameld. Een belangrijke stap in de formulering van ontwerpcriteria is het formuleren van structuur-activiteitsrelaties die erop gericht zijn om katalytische prestaties te relateren aan typische katalysatoreigenschappen. Daarvoor is de katalytische activiteit bepaald van ontzwaveling in de gasfase van dibenzothiopheen.

In hoofdstuk 3 is de invloed onderzocht van de inzwavelingsdruk op de structuur en activiteit van gecalcineerde $CoMo/Al_2O_3$ katalysatoren. De hoeveelheid Co atomen die participeert in de actieve fase is hoger na inzwaveling bij 4 MPa dan na inzwaveling bij atmosferische druk. Een studie van de stapsgewijze inzwaveling wijst op hogere inzwavelingssnelheden van Co en Mo bij verhoogde druk. Dit leidt tot een vorm van bulk CoS_x deeltjes bij tussenliggende temperaturen, gevolgd door een redispersie van deze deeltjes naar kleinere deeltjes. Deze redispersie gebeurt op het moment dat MoS_2 *slabs* worden gevormd. Verder wordt aangetoond dat een typisch gecalcineerd $CoMo/Al_2O_3$ monster, dat zich als een Type I katalysator gedraagt na atmosferische inzwaveling, de veel actievere Type II katalysator oplevert na een hoge-drukbehandeling. Deze waarneming werpt de vraag op of en in welke mate het mogelijk is om Type I Co-Mo-S actieve fasen onder industriële inzwavelcondities te behouden.

Om de invloed van de condities bij calcinering op de prestatie van $CoMo/Al_2O_3$ katalysatoren beter te begrijpen zijn in hoofdstuk 4 een aantal ongecalcineerde katalysatoren met verschillende Co/Mo verhoudingen onderzocht. Wanneer de calcineerstap wordt weggelaten nadat de gecalcineerde Mo/Al_2O_3 monsters zijn geïmpregneerd met Co resulteert een verminderde interactie van de Co-ionen met de alumina drager. Daardoor zullen alle Co-atomen reeds volledig ingezwaveld zijn bij lagere temperaturen. Als de Co/Mo-verhouding relatief hoog is, zal een belangrijke fractie kleine CoS_x deeltjes gevormd worden of hoog gedispergeerde Co_9S_8-achtige deeltjes met een lage activiteit. Hoewel de CoS_x-structuur MES parameters kan opleveren die lijken op die van de Co-Mo-S-fase, kan MES toch de aanwezigheid aantonen van de verschillende Co-sulfide soorten omdat kleine quadrupole splitsingen (QS) wijzen op de aanwezigheid van grotere Co-sulfide deeltjes. De onder hoge druk ingezwavelde monsters hebben een veel hogere HDS activiteit dan katalysatoren geactiveerd bij 0.1 MPa. Het verschil wordt veroorzaakt door de vorming van Type II Co-Mo-S bij hogere drukken. De activiteiten van de monsters met Co/Mo-verhoudingen van 0.30 en 0.52 zijn vergelijkbaar, hetgeen de conclusie ondersteunt dat katalysatoren met meer Co een belangrijk deel afgescheiden CoS_x-deeltjes zal opleveren.

Hoofdstuk 5 evalueert de toepasbaarheid van het Co-Mo-S-model en de stabiliteit van de Co-bevattende fases voor katalysatoren op een koolstofdrager onder hogedruk inzwavelcondities. De zwakke interactie van de Co-atomen met het dragermateriaal en ook de zwakke Co-Mo-interactie leidt tot een aanzienlijke vorming van bulk-achtige Co-sulfide deeltjes na inzwaveling bij 0.1 MPa. MoS_2 verhindert complete sintering van de CoS_x-structuren onder atmosferische druk, maar bij inzwaveling op 4 MPa sinteren deze CoS_x-deeltjes naar Co_9S_8-achtige structuren. Inzwaveling bij atmosferische druk zorgt voor actieve deeltjes, waarschijnlijk hoog gedispergeerd Type II Co-Mo-S gevangen in de microporiën van de koolstofdrager. De lage toegankelijkheid heeft een negatief effect op de

DBT HDS activiteit, maar de tiophene HDS activiteit is relatief hoger. Echter, inzwaveling op 4 MPa leidt tot sintering van de MoS_2-fase hetgeen een lagere tiopheen HDS activiteit laat zien. De DBT HDS activiteit is nu echter veel hoger omdat de deeltjes niet langer in de microporiën gevangen zitten.

Om de invloed van hoge druk inzwaveling van $CoMo/Al_2O_3$-katalysatoren die bereid zijn langs niet-conventionele wegen beter te begrijpen, zijn NTA-houdende (nitrilotriazijnzuur) en fosforhoudende CoMo-katalysatoren onderzocht in hoofdstuk 6. De aanwezigheid van toevoegingen die de interactie van metaal en drager verzwakken, leiden tot een aanzienlijke vorming van CoS_x-deeltjes. In de katalysatoren met lage Co-belading zijn alleen Co-Mo-S-structuren aanwezig, maar bij de katalysator die met een grotere hoeveelheid NTA werd bereid, werd een verbeterde dispersie waargenomen. De spectroscopische gegevens suggereren dat complete vorming van Type II Co-Mo-S alleen plaatsvindt na behandeling onder hoge druk. Katalysatoren die behandeld waren op 4 MPa gaven hogere activiteiten te zien dan degene geactiveerd bij atmosferische druk.

In hoofdstuk 7 wordt getracht om de sterk variërende DBT HDS activiteiten uit eerdere hoofdstukken te begrijpen. Omdat de spectroscopische resultaten, die de Co-omgeving bekijken (MES, EXAFS) geen significante verschillen in de eindtoestand aanduiden is de aandacht vooral gericht op het Type I/II gedrag van de actieve fase. De aanwezigheid van een continuum van Co-Mo-S-structuren vanaf het weinig actieve Type I met een sterke dragerinteractie tot een volledig ingezwaveld en mogelijk kristallijn Type II Co-Mo-S met een zwakke dragerinteractie wordt voorgesteld voor ingezwavelde CoMo katalysatoren. Het gebruik van NTA en hogedrukinzwaveling leidt tot het sterkste Type II karakter van de actieve Co-Mo-S fase. Een redelijke indicator voor het Type I/II karakter van de actieve fase lijkt het Mo-S coördinatiegetal te zijn, verkregen middels Mo K-edge *X-ray absorption spectroscopy*. Tegelijk met het bereiken van een hoge inzwavelingsgraad in de NTA-katalysatoren treedt het stapelen van de MoS_2 *slabs* op. Dit stapelgedrag lijkt weinig invloed te hebben op de gasfase tiopheen ontzwaveling, maar hindert in zekere mate de gasfase DBT ontzwaveling. Er wordt voorgesteld dat de DBT HDS activiteit van de op hoge druk ingezwavelde CoMo-NTA-katalysator (pure fase II) met een enigszins gestapelde MoS_2 morfologie verbeterd kan worden door het stapelen van de actieve fase te beperken. De meest actieve katalysator in DBT HDS is een op 4 MPa ingezwavelde gecalcineerde CoMo katalysator. De hoge activiteit wordt verklaard door een bijna complete inzwaveling naar Type II terwijl de morfologie van de MoS_2 fase bijna die van een monolaag is. De voorwaarden van een lage stapelgraad en volledige inzwaveling van MoS_2 worden bediscussieerd voor hoog-actieve CoMo-sulfide katalysatoren.

Hoofdstuk 8 beschrijft voor de eerste keer het gebruik van ^{182}W Mössbauer spectroscopy om de inzwaveling van Al_2O_3- en amorfe-silica-alumina-gedragen NiW-katalysatoren onder industriële condities te bestuderen. ^{182}W MAS kan een duidelijk onderscheid maken tussen WO_3- en WS_2-achtige fases die men tegenkomt in de gecalcineerde en ingezwavelde vormen van op W gebaseerde *hydrotreating* katalysatoren. Behalve dat de inzwavelgraad bestudeerd kan worden, geeft ^{182}W MAS ook een schatting van de mate van kristalliniteit van de WS_2 deeltjes, hetgeen relevant is voor fase I/II gedrag in NiW-katalysatoren. Deze informatie is zeer bruikbaar, maar kan ook directer verkregen worden met meer conventionele technieken. Voor NiW gedragen op alumina en amorfe silica alumina werd waargenomen dat verbeterde dragerinteractie door verhoogde calcineringstemperatuur tot een moeilijkere inzwaveling van W leidt. In de gevallen waar de inzwavelingssnelheid het laagst is, is een tussenliggende WS_3-achtige fase geidentificeerd bij temperaturen rond 673 K. Ni zorgt voor een betere inzwaveling en in dit geval gebeurt de inzwaveling via een oxysulfidische tussenvorm. Een algemene conclusie is dat inzwaveling bij atmosferische druk leidt tot een slecht kristallijne WS_2-fase, terwijl inzwaveling op 4.0 MPa de kristalliniteit verhoogt, hetgeen bewezen worden door de benadering van de spectrale parameters van die van WS_2.

In deze studie zijn twee belangrijke factoren geïdentificeerd die de HDS-activiteiten van CoMo *hydrotreating* katalysatoren beïnvloeden: de vorming van CoS_x-structuren in katalysatoren met een zwakke interactie tussen metaal en drager en de aanwezigheid van een continuüm van Type I/II Co-Mo-S-structuren. Het werd bevestigd dat MES een nuttige techniek is om op zijn minst het aantal promoter atomen in Co-sulfide deeltjes aan te geven bij een variable aantal Co atomen. De voorgestelde richtingen voor toekomstig onderzoek zijn: (i) een beter begrip van het stapeleffect op de HDS activiteit en (ii) de bepaling van de beste bereidings- en activeringsmethoden om MoS_2 volledig in te zwavelen met een redelijkerwijs hoge dispersie. Dit proefschrift toont ondubbelzinnig het belang aan om verdere experimenten uit te voeren onder voor de industrie relevante temperaturen en drukken. Een betere kwantitatieve beschrijving van de Co-deeltjes aanwezig in Co-sulfide katalysatoren, specifiek door MES spectra te deconvolueren in verdelingen van Co-sulfide deeltjes van verschillende grootte, wordt voorgesteld als een nuttige aanpak in toekomstig onderzoek.

Publications and presentations

Publications:

A.I. Dugulan, M.W.J. Crajé and G.J. Kearley, *High-pressure in situ Mössbauer emission spectroscopy study of the sulfidation of calcined CoMo/*Al_2O_3 *hydrodesulfurization catalysts*, Journal of Catalysis 222 (2004) 281.

A.I. Dugulan, M.W.J. Crajé, A.R. Overweg and G.J. Kearley, *The evolution of the active phase in CoMo/C hydrodesulfurization catalysts under industrial conditions: a high-pressure Mössbauer emission spectroscopy study*, Journal of Catalysis 229 (2005) 276.

A.I. Dugulan, M.W.J. Crajé, A.R. Overweg and G.J. Kearley, *Evolution of the active phase of CoMo/*Al_2O_3 *catalysts under industrial conditions: a high-pressure MES study*, AIP Conference Proceedings 765 (2005) 26.

A.I. Dugulan, E.J.M. Hensen, J.A.R. van Veen, *High-pressure sulfidation of a calcined CoMo/*Al_2O_3 *hydrodesulfurization catalyst*, Catalysis Today 130 (2008) 126.

Chapters 4 through 8 of this thesis will be submitted for publication.

Presentations at conferences:

Mössbauer emission spectroscopy study of the high-pressure sulfidation of calcined CoMo/Al_2O_3 *hydrodesulfurization catalysts*, Netherlands' Catalysis and Chemistry Conference - NCCC V, March 2004, Noordwijkerhout, The Netherlands.

Elucidation of the Co phases present in CoMo hydrodesulfurization catalysts sulfided under industrial conditions: a high-pressure Mössbauer study, Molecular Aspects of Catalysis by Sulfides - MACS III, May 2004, Ascona, Switzerland.

Evolution of the active phase of CoMo/Al_2O_3 *and CoMo/C catalysts under industrial conditions: a high-pressure MES study*, International Symposium on the Industrial Applications of the Mössbauer Effect - ISIAME, October 2004, Madrid, Spain.

Effect of pressure on the genesis of the Co-Mo-S phase in CoMo/Al_2O_3 *catalysts: a Mössbauer emission spectroscopy study*, Netherlands' Catalysis and Chemistry Conference - NCCC VI, March 2005, Noordwijkerhout, The Netherlands.

^{182}W *Mössbauer spectroscopy study of the sulfidation behavior of NiW/*Al_2O_3 *hydrodesulfurization catalysts*, International Conference on the Applications of Mössbauer the Effect - ICAME, September 2005, Montpellier, France.

Sulfidation behavior of CoMo/Al_2O_3 *and NiW/*Al_2O_3 *catalysts under high-pressure conditions:* ^{57}Co *MES and* ^{182}W *MAS studies*, Molecular Aspects of Catalysis by Sulfides - MACS IV, May 2007, Doorn, The Netherlands.

Acknowledgements

My last remaining task before finishing this PhD thesis is to acknowledge all those people that have contributed to the work described here. To all of them, I am truly indebted.

First of all, I would like to express my deepest gratitude to my *promotor* and supervisor, Prof. Dr. Rob van Veen, for his continued encouragement and invaluable suggestions during the final years of this study. Without his enthusiastic support and contributions, the completion of this thesis would not have been possible. Secondly, I am extremely grateful to Dr. Emiel Hensen for his guidance and inspiration. Dear Emiel, I will never forget how many good ideas you offered me and how you helped in solving experimental problems. Without you, there would have been no EXAFS data in this thesis. I also thank you for giving me access and the necessary support to perform catalytic activity measurements, and for your tireless effort to correct all the chapters of the thesis. I am very much thankful to my previous supervisors Dr. Menno Crajé and Dr. Arian Overweg for giving me opportunity to pursue my PhD study here at Delft University of Technology and for their valued suggestions and insightful ideas.

To Prof. Dr. Ignatz de Schepper I extend my warmest thanks for accepting the responsibility to by my *promotor* and for his kind assistance during the finalization of this dissertation. Prof. Dr. Gordon Kearley and Dr. Adri van der Kraan are sincerely acknowledged for their expert advice and guidance. Dr. Patricia Kooyman is gratefully appreciated for acquiring the TEM images and for the time spent for fruitful scientific discussions. I am also deeply indebted to Dr. Menno Blaauw for his generous support and encouragement.

Very special gratitude goes to Michel Steenvoorden for his skillful technical assistance in performing the experiments (*i.e.* Mössbauer, XAS), for all the help with the old and the new Mössbauer software, and for the Dutch version of the summary and propositions. His friendship and the nice conversations outside science kept me going through the toughest parts. I would also like to thank Wouter Borghols and Dilip Poduval (TU/e) for the support and cooperation given during the XAS experiments. Dr. Sergey Nikitenko (ESRF, France) is kindly appreciated for his time and help at the *DUBBLE* station. My sincere thanks go also to Robert van Teeffelen (TU/e) for the technical support during the catalytic activity measurements. I am greatly indebted to René Gommers and René Nouse for the design and manufacture of the tungsten *in situ* cell.

I am thankful to all my colleagues and friends at IRI (RID) for discussions and comments during the group meetings and *gezelligheid* during the lunch and coffee breaks. A special mention to my roommates: Jerome Taboada, Serdar Sakarya, Yulia Efimova, and Sarita Singh; thanks a lot for the support, friendship and endless good mood. To my Romanian friends Cãlin Zamfirescu, Daniela Neamțu, Ramona Apostol, and Cosmina Druicã, thanks for sharing with me the joy and pain, frustration and happiness and for encouraging and helping me in any possible way.

My unconditional gratitude and love goes to my family and closest friends, especially to my mother and my brother Adrian. Otilia, you are the reason I got involved in research, and although you are far away now, our affection will never end. Ella, your appetite for life and the pleasure you took in making me a part of it gave me the necessary incentive to complete this work.

Curriculum Vitae

Achim Iulian Dugulan was born on the 11th of February 1977 in Râmnicu Vâlcea, Romania. There he attended the primary school and in June 1995 he graduated from *Tudor Vladimirescu* high school in Craiova, Romania. In September 1995 he continued his education in the Faculty of Physics at University of Bucharest. He received his Bachelor of Science with specialization in Physics Engineering in June 2000. In 1996 he also started studying Chemical Engineering at Technical Academy of Bucharest, from which he earned his Bachelor's degree in June 2001 and his Master of Science degree in June 2002. In August 2002, he started his PhD project at the section Fundamental Aspects of Materials and Energy, Department of Radiation, Radionuclides & Reactors, Faculty of Applied Sciences, Delft University of Technology. The significant results of his PhD research are described in this thesis. Since February 2007 he works as a researcher at Reactor Institute Delft (RID), Delft University of Technology.

